The Programmable Logic Controller

Learn how to Program & Troubleshoot Ladder Logic

By Curtis Green

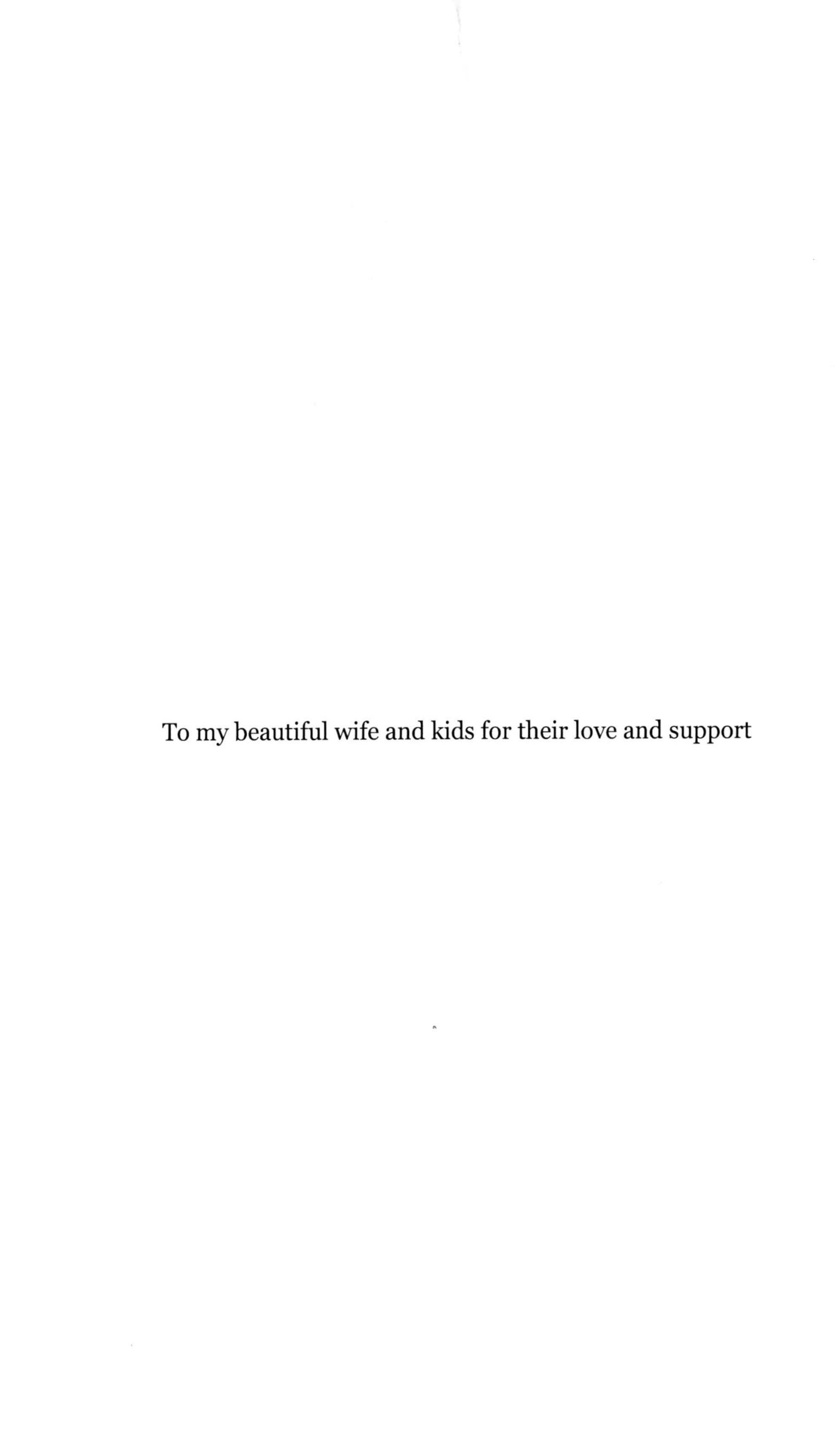

To my beautiful wife and kids for their love and support

Table of Contents

(This Page intentionally left blank)

Introduction

Everywhere in the modern world from Anaheim to Hong Kong a revolution is in full swing. Technologies of the past are systematically being replaced by automated systems that sort, ship, manufacture, assemble and process goods.

Have you ever wondered what happens to your luggage after you check it in at the airport? What mechanism allows your luggage to be screened and delivered to your plane quickly and efficiently? What about online outlets like Amazon? What level of automation is in action behind the scenes? What enables the transfer of goods from a warehouse across the country to your doorstep in less time than ever before? How are manufacturing plants able to produce millions of widgets quickly with minimum labor and maximum quality?

Automation is at the heart of modern industry. It is predominant in distribution, processing, manufacturing, and assembly. Behind the robots and machinery lies the Programmable Logic Controller or PLC for short. Open any modern control panel and you will find the programmable logic controller has replaced an array of relays, timers and other antiquated devices.

The contemporary PLC dominates the world of industrial automation and Ladder Logic is the de-facto programming language in this obscure land of bits and bytes. Ladder Logic used to be relegated to one-line diagrams in black and white, residing on paper tucked away in the doors of control panels. Technicians would use them for troubleshooting when things went wrong, as they often did. They were used to describe how all the relays, timers, contacts, sensors and coils intermingled to create logical circuits. Today these one-line diagrams have grown into a dynamic programming language we call Ladder Logic. Ladder Logic, Function Block Diagram and Sequential Function Chart are the graphical programing languages defined by the IEC 61131 standard for programmable logic controllers. As it is the most widely used, we will solely be focusing on Ladder Logic.

This automation revolution has created a need for engineers and technicians that can program, design, commission and maintain automated control systems. As of the writing of this book the demand for qualified people far outweighs the supply, which for you, the future automation engineer, means a rewarding career. Put simply you can have a better job with better pay. The automation world is in need of skilled individuals who know automation, specifically programmable logic controllers, industrial circuits, robots, vision systems and SCADA systems.

Your free gift.

I created a manual that details the Graphical User Interface of the Logix5000 programming and configuration software. This free manual compliments this book. The manual covers key areas such as the controller organizer, tag editor and IO Configuration tree. It's free and exclusive to those who have purchased my book.
www.ladder-logic.com/book-free-gift/

www.ladder-logic.com/book-free-gift/

C-1 What is a PLC and Why Ladder Logic?

The Programmable Logic Controller, or PLC is an industrial computer that lies at the heart of modern day automated systems. Its purpose is to control machinery which include assembly lines, manufacturing cells, waste water treatment plants, and material handling systems to name a few.

Most PLC's are modular, they can be scaled to fit any desired application. Input and Output devices can be added and removed as necessary. IO modules are available in both discrete and analog models. Other module types are available for communications, robotics, motion control, and vision. Specialty modules are available that can run DOS or Windows-based operating systems and can be programmed to do any number of tasks.

PLC's also come in small, compact containers for use on projects requiring small programs and just a few IO points. Most PLC programs run continuously, scanning inputs, executing logic and activating outputs, however programs can be configured to run periodically at predetermined time intervals or triggered by events.

Inputs and Outputs

Inputs can be either discrete or analog. Discrete Inputs have two states, *on* and *off*. Inputs can also be variable, such as a temperature input. These inputs are known as Analog Inputs and can have multiple values within a predetermined range. Outputs are similar to inputs; there are Discrete Outputs and Analog Outputs. A Digital Output might be a simple pilot light, while an Analog Output might be a variable current or voltage.

PLC's are programmed with specialized software residing on a programming laptop or desktop PC. They are typically programmed using a language called Ladder Logic, however they can be programmed using Structured Text, Function Block, or Sequential Function Chart, with Ladder Logic being the most common control language.

A PC can download to, upload from, or go online with a PLC. Communication capabilities vary in regards to the manufacture and model of the controller. Some of the more common communication methods are Serial, USB, and Ethernet. There are many, many more; it all depends on the architecture of the control system.

What is Ladder Logic?

So what is all this talk about Ladder Logic, what is it? Why Ladder Logic? Ladder Logic is a language used to program PLC's. Industrial automation has a language all its own and has been dubbed Ladder Logic. Interestingly enough it is one of the most misunderstood and curious languages out there. This is probably the case for several reasons. The language is graphical, not textual like Java or C++, it mimics one-line diagrams which are electrical in nature. It was made for electrical folks, not your typical programmer cranking out functions and classes for web and gaming applications.

Picture if you will a step ladder. Each step on the ladder is referred to as a rung. The rungs of the ladder are attached to rails. Ladder Logic is laid out the same way. The rails represent the opposing polarities of an electrical potential and the rungs represent the connections between the two.

Rather than writing code in a text editor, rungs are placed between rails and instructions are placed on rungs. The rungs along with the instructions and variables make up the language of Ladder Logic. Ladder Logic instructions were developed to mimic electromechanical components from the not-so-distant past. The software is intended to look and act like these industrial circuits.

Why This Book is For You

My introduction to automation came not from school, but from work which I found is often the case. I worked nights as a maintenance apprentice at a dairy processing plant while attending school during the day. At the heart of this operation was an old Westinghouse PLC. Getting online with the old beast at least once a day to troubleshoot or force coils on or off was an inevitability. As the operations expanded, smaller, more sophisticated programmable logic controllers replaced the old bulky Westinghouse controller. Standalone PLC's were replacing relay logic, timer modules and remote IO modules. Larger machinery with multiple processes were replaced by compact scalable SLC -500 controllers. The SLC family of controllers predate the Logix5000 family of controllers. Even before the SLC-500 there were PLC-5's.

As this new technology became more prevalent it was clear there was a lack of expertise at the plant, there were no engineers or qualified technicians. It turned out that our local self-proclaimed "guru" wasn't much of a guru at all.

As the new controls equipment was being installed a controls engineer set up a PC that could communicate with all the connected devices, it was amazing. The only problem was that none of us knew what we were doing. We could force an input on to jumpstart a process as needed, but none of us knew why the stupid thing wouldn't come on when it was supposed to. The plant was quickly becoming fully automated, we all needed to step up our game.

The night shift was relatively slow and afforded me the opportunity to study the language, to find out what was really going on inside the magic box. I started learning on a little

Micrologix I borrowed from the spares department. All I had to do was hook it up to a power supply and get online with it, which I eventually did. Once online I tried furiously to write some code. All I wanted to do was get a simple flasher circuit to work. I wanted to turn an output on and off. I spent a considerable amount of time trying to make a circuit work. After much frustration I eventually accomplished my task.

I was hooked. I read every manual I could get my hands on. The manuals were technical in nature. I would take them home and read them before bed, (incidentally if you have insomnia it's a great way to fall asleep). What I found is that it was not a good way to learn something that is, in fact, fairly simple. It's not that the publications I was reading were not helpful; they were packed full of good, solid technical information. I could tell you the most minuscule detail of an instruction, how many bits were consumed, how much scan time a given instruction would use or what the pre-scan behavior was. The problem was that none of this information could be put to use in a practical sense. I still didn't know how to program or how the logic worked. I could identify various instructions in a program, but I didn't yet know how to put it all together.

By this time I knew every instruction available, but I didn't know how to implement any of them. I kept at it and eventually befriended the controls engineer who was installing the new automation equipment. I asked him many questions which he unfailingly answered with patience and generosity. This was considerably better than reading technical publications. This guy actually showed me how the logic worked. After a substantial amount of guidance from my mentor and even more practice I was finally ready to have a go at it myself.

I wrote this book for myself, or rather for the would-be programmer from years ago. This book covers the basic elements that will accelerate the learning curve for someone who is starting the journey of learning how to program Ladder Logic. I wish this book had existed years ago as it would have saved me quite a bit of time and frustration. I could have hit the ground running rather than stumbling and finding my way on my own. This is a compression of all the knowledge I gleaned from countless manuals, publications and discussions with controls engineers. I wrote this book to give you the knowledge that I spent years looking for and eventually painstakingly obtained.

What To Expect

To you, the future controls engineer. Let this be your resource, use it to its full potential. Long before you need to know all the instructions, languages and architecture of PLC's you need to understand the logic. You need to know how to read and write Ladder Logic. You should be able to open any program and read it like you are reading this book.

I feel it's my obligation to point out that this book is not for everyone. I won't be talking about every instruction. I won't be telling you the same stuff you can get by opening up the help files that come with the software. As an example, if you want to know exactly how an FSC or BTD instruction works you are not going to find it here as that technical information is readily available elsewhere. This book is intended to lay a foundation, give you a solid start. If you read the book and practice the principals laid out you will be able to open any Ladder Logic program and understand it. You will have enough knowledge to be able to successfully troubleshoot projects, modify them and even create projects using basic instructions and techniques. You will be able to identify tasks and programs. You will be able to identify all the various tags including structures and arrays. You will understand why the logic is organized in tasks, programs and routines. You will also be able to create programs using all the components mentioned. This ***is*** the place to start.

It is important to note that there are several PLC manufactures, each with their own proprietary version of Ladder Logic. This book makes use of examples inherent to RSLogix 5000 Programming and Configuration software from Allen Bradley. There are as many versions of Ladder Logic as there are manufacturers such as Siemens, GE, and Fanuc to name a few. Every version of PLC has a different version of Ladder Logic. It would be convenient if the logic were the same across every platform, however the reality is that it is not.

The Good News

The good news is that if you learn one platform the knowledge can easily be used to learn others. The instruction sets are similar and there are more commonalities than differences. Every version has specifics to its platform however, if you learn one version of Ladder Logic you are 80 percent of the way to learning all the others.

Did you know that there is an IEC standard for PLC programming languages? RSLogix 5000 software complies with this standard making it an excellent platform to learn on and it is also one of the most popular versions in use today.

Although not necessary to complete this book, eventually you will need access to the programming software, this book alone will not make you a programmer, it will take work and practice so let's get started.

C-2 Ladder Logic Instructions, Some Basics

Ladder logic is made up of instructions and variables. Most instructions require at least one variable which is referred to as the instructions argument or operand. Nothing like referring to the same object with different names to complicate things. To make things simple think of instructions as either input or output. Input instructions examine data while output instructions write data.

Basic Ladder Logic instructions are referred to as Boolean instructions. According to Wikipedia: "In computer science, the Boolean data type is a data type, having two values (usually denoted true and false), intended to represent the truth values of logic and Boolean algebra. The Boolean data type is the primary result of conditional statements, which allow different actions and change control flow depending on whether a programmer-specified Boolean condition evaluates to true or false." What? To simplify, in ladder logic a Boolean input instruction, when solved, evaluates to either true or false.

Boolean Input Instructions

To better understand instructions let's examine a basic input instruction named Examine If Closed (XIC). Think of it like a switch, when the switch is closed the program scan is allowed to pass to the right of the instruction. When the switch is open everything to the right of the instruction is inaccessible to the program scan. The instruction examines the value of its associated argument. Slightly confusing to the beginner, isn't it?

The argument refers to a value, for now consider the value to be either 1 or 0 also known as true or false respectively. If the value is true the "switch" is closed and open if the value is false. Again, an argument that contains the value of 1 is considered true and false when 0.

Compare this to the common household light switch. When the light switch is closed a path is created allowing electrons to flow through the light bulb filament thereby illuminating the room. Let there be light! When the switch is turned off the path to the light is removed. Hello, Dark Ages! The XIC instruction behaves like the light switch, when it is closed a path is created and when it is open the path is removed.

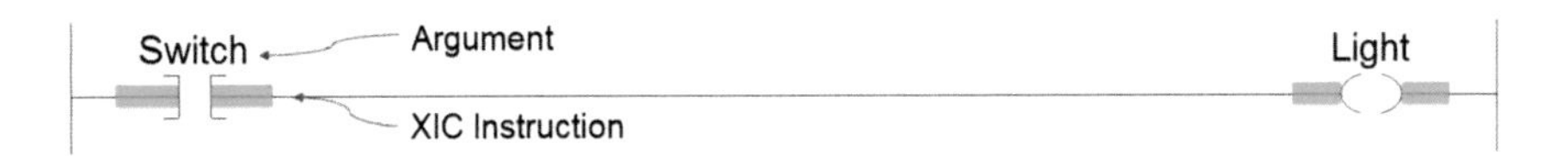

Input instructions examine a value and make a decision based on that value. The XIC instruction is considered a BOOLEAN input instruction. The XIC instruction is a symbol representing a normally open contact or switch. The instruction evaluates the argument which will be either true or false. The argument is a value or, rather, an address to a value. The following examples show two identical rungs. The value of the XIC argument of the first rung is false and the value of the second XIC argument is true. Another way of looking at this could be that in the first rung the switch is open and the light is off. In the second rung the switch is closed and the light is on. Notice that the instructions are highlighted in green when true and not highlighted when false.

If the XIC instruction represents a normally open switch then the XIO instruction (Examine If Open) represents a normally closed switch. The XIO instruction evaluates as true when the value of the argument is 0 and false when 1. It is the opposite of the XIC instruction, convenient don't you think? Using the same argument with a different instruction yields different results. What if you were programming a system and you wanted to turn on a prompt light when a switch was not made?

Boolean Output Instructions

Thus far we have talked about two basic input instructions, but what about the output instruction on the rungs above? The Output Energize instruction (OTE) is also a Boolean instruction, however it is considered an output instruction. Instead of evaluating the value of the argument it sets the value. If the logic before the instruction evaluates as true then it sets its argument to true. If the logic evaluates as false the argument is set to false.

Ladder Logic was written to emulate electrical ladder diagrams. The basic components of ladder diagrams are contacts and coils. Thinking about ladder logic in terms of contacts and coils is an essential part of understanding the language. Contacts and coils are symbols that can be placed on rungs to create logical circuits.

When the Input Contact (XIC instruction) is closed a path is created allowing "current" to flow to the Output Coil (OTE instruction). In a real electrical circuit the contact might be a mechanical switch or button actuated manually. The action of closing the switch allows electrons to flow through the circuit resulting in the output coil turning on. The output coil could be a light, horn, motor-starter, etc.

The coil symbol is called an *Output Energize* or OTE instruction. The OTE instruction also requires an argument similar to the XIC instruction. Rather than checking the value of the argument the OTE sets the value of the argument to either a one or zero. If the XIC instruction is true a path is created and the OTE sets the associated argument to true.

Input instructions *check* values of associated arguments, while output instructions *set* values of associated arguments. Arguments can also be called Operands. The most commonly used instructions are Boolean instructions. Boolean instructions have two possible states which can be referred to as one and zero, on and off, set and clear, and true & false.

Timer & Counter Instructions

Not all instructions are Boolean in nature. As an example, timer and counter instructions are considered output instructions and operate with structured arguments, which we will explore later. As the name implies a timer instruction is used to track time. If you need a light to turn on for two seconds then turn off you could use a timer instruction. There are several timer instructions: Timer-On (TON), Timer-Off (TOF) and Retentive-Timer (RTO). The TON begins timing when the rung input evaluates as true. The TOF begins timing when the rung input evaluates as false, and the RTO instruction pauses when the rung input evaluates as false.

The counter instructions simply count up or count down depending on which instruction you use. The Count-Up (CTU) instruction will increment a count on a false to true transition. The Count-Down (CTD) instruction will decrement a count on a false to true transition. These instructions can be used independent of each other or together to increment and decrement a number. Just like timers, counters can be used to solve any number of problems. These are great for counting product going in and going out. They can also be used to count machine cycles as well as anything else you can come up with.

Copy & Move Instructions

In many cases data will need to be copied from one place to another. The two main instructions for copying data are the Copy and Move (COP, MOV) instructions. They both copy data from one address to another. The main difference is the MOV instruction is intended to copy one element while the COP instruction will copy multiple elements like a string. A string is an array, which has multiple elements.

Simple Math Instructions

There will be times where your program will need to do math, for instance I recently worked on a leak-check machine. The leak-check machine floods one side of a production part with helium while pulling a vacuum on the other side of the part. The vacuum side has a mass-spectrometer which is calibrated to detect how much helium passes through a part. The mass-spectrometer would then send the leak rate data to the PLC via an analog voltage signal.

The program I created needed to convert this voltage into a helium leak rate, which was accomplished using basic math instructions. The program then compared the leak rate with known leak rate values. If the leak rate was out of the range of the known values the part was failed and scrapped, if the part fell between the known values the part was given a pass status.

Math instructions are considered output instructions and are usually preceded by input conditions, although not necessary. There are Add, Subtract, Multiply and Divide instructions. The Compute (CPT) instruction is handy instruction that allows you to use operators to build equations. For instance the equation (45-23)/2 could be placed in the CPT instruction as opposed to using separate Subtract and Divide instructions.

Compare Instructions

Compare instructions are considered input instructions that compare values. The equal or EQU instruction compares two values, if they are equal the input is considered true and the output instruction is enabled.

There are several comparison instructions available. Consider the Not-Equal (NEQ) instruction, if the two values are equal the input is considered false and the output is disabled, however when the two values are not equal the input is considered true and the output is enabled. Other comparison instructions are the GRT, GEQ, LES and LEQ. Greater Than, Greater or Equal to, Les Than and Less Than or Equal to.

Program Flow Instructions

Program flow can be altered with instructions like the jump to label combination JMP & LBL. When the program scan reaches a JMP instruction the scan is transferred to the LBL instruction. This can be used to skip logic or create a conditional loop.

The Jump to Subroutine or JSR instruction will transfer the program scan from the current routine to a different routine in the program.

Summary

These instructions are just the tip of the iceberg, they are simple yet powerful and used more than all of the other instructions combined. There are a variety of instructions to manage communications, files, sequences, motions, math, etc. Most programs use only a handful of instructions. It would be a disservice to write a book that explains every instruction, as this is covered in the help files of the programming software quite extensively.

End of Chapter Review

At the end of most chapters there are a few questions followed by answers. You should be able to get all of the answers correct, if not review the chapter again in order to ensure that you have fully grasped all of the outlined concepts. I recommend you do the tests as they are designed to reinforce the information you are learning as you go through the book.

Chapter 2 Questions

1. Most instructions require an argument. An argument refers to a _____________?

- Tag
- Pointer
- Module
- Something you do with your spouse.

2. What does OTE stand for?

- On Then Examine
- Output Energize
- Output Enable
- Off Timer Enable

3. The XIC instruction can be considered an _________.

- Instruction of program scope.
- Output instruction.
- Input Instruction.
- Instruction of varying arguments.

4. True or false, a Boolean data type can hold and store integer numbers up to 7.

5. True or false, an operand is another way of saying Boolean.

6. Instrcutions are symbols that are placed on rungs. When the program scan reaches an instruction an action is taken. When a rung is scanned it is always scanned from ____________.

- left to right.
- right to left.

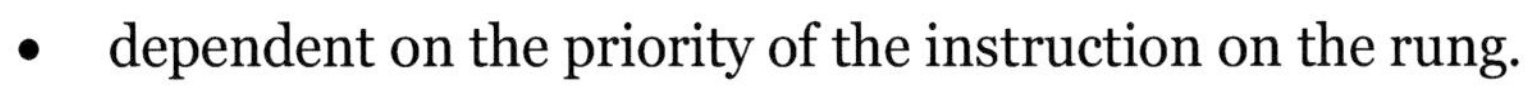

- dependent on the priority of the instruction on the rung.
- a configurable starting point.

Chapter 2 Answers

1. Most instructions require an argument. An argument refers to a tag.

   A tag is an address to a location in the PLC'c memory.

2. The OTE instruction stand for?

   Output Energize. The OTE instruction is an output instruction that will change the value of the associated argument or tag.

3. The XIC instruction can be considered an input instruction.

   Input instruction. The instruction will not modify the value addressed by the argument.

4. True or false, a Boolean data type can store integer numbers up to 7.

   False, a Boolean data type can store either 1 or 0, also known as true or false respectively.

5. An operand is another way of saying Boolean.

   False. The term operand refers to an instructions argument. Arguments, or operands refer to tags.

6. Instructions are symbols that are placed on rungs. When the program scan reaches an instruction an action is taken. When a rung is scanned it is always scanned from left to right.

   Left to right. The scan of a rung is always from left to right. Instructions on the left are scanned before instructions on the right. This is also true for top to bottom.

C-3 Variables

Definition

Before you can fully understand how a program works you must first understand and identify the basic components of a program. All programing languages make use of variables. Ladder-logic, C++ and Java all use them. A variable according to Wikipedia "is a storage location paired with an associated symbolic name (an identifier), which contains some known or unknown quantity or information referred to as a value." In the controls world variables are referred to as tags.

Tags Are Variables

A tag meets all the criteria called out by the definition above. It has a symbolic name which is the identifier and contains a value. You should recognize tags as the arguments for instructions. In the previous chapter the XIC, XIO and OTE instructions all had arguments named SWITCH and LIGHT.

Before a tag can be used with an instruction a strict definition is required. In order to do that we must first know exactly what a tag is. A tag is an address to a memory location. Somewhere in the depths of the PLC exists a vast array of empty memory banks waiting to be filled with data. Accessing these memory locations is easy; all you need to do is create an address. You do this by defining a tag, the name then becomes the address. You also define how much memory will be allocated to this address and how the data will be stored. The software automatically associates your new tag name to a spot in memory. To access the memory you simply use the tag's name. Below is a screenshot of the properties dialog box of a tag named "Door Closed". The tag has been defined as a BOOL which is short for Boolean and means the tag will consume one bit of data.

Tag Data Types

Not only does a tag require a descriptive name it also must define how much memory the tag will address. Will the tag reference a single bit of data or will it occupy several bits? Will the tag hold a number or a string?

Three bits are required to store the number seven, and four bits are required to store the number sixteen. Rather than creating a custom data size for every number there are several native data types available referred to as Atomic data types. All other defined data types originate in one way or another from these basic five native data types:

- BOOL 1 bit
- SINT 8 bits
- INT 16 bits
- DINT 32 bits
- REAL 32 bits

A data type then refers to how much memory a tag will consume. For example defining a tag with the name MY_TAG of data type DINT will reserve 32 bits of memory addressable by the name MY_TAG, and store a decimal number with a value between -2,147,483,648 and 2,147,483,647. A tag defined with the name MY_BOOL with the data type BOOLEAN can hold a value of 1 or

0. A tag defined with the name MY_STRING with the data type STRING can hold 82 characters. The STRING data type is a pre-defined data type consisting of an array of 82 SINT's and one DINT.

Creating Tag Names

Tags are named when they are defined. Simple enough, right? There are a few governing rules that must be followed when naming tags. Tags are a minimum of one character and must start with an underscore'_' or Alpha character 'A-Z'. Both lower case and upper case are supported and are considered equal. 'A' is the same as 'a'. As a general rule tag names should be descriptive, the name should describe what the tag will be used for.

For instance, suppose you are writing a program that requires a door to be closed in order for a motor to run. Your job is to give descriptive names to tags. What would you call them? You could name it after your nephew, girlfriend, boyfriend, or dog, but what if you gave it a descriptive name that makes your code legible? Look at the following two examples, which one is more easily understood; rung one or rung two?

Obviously the second rung makes much more sense, it tells a story. If the door is closed start the motor. Notice that rung one is the exact same logic as Rung two, however what is the objective of rung one? Why did the programmer write the rung? The logic is the same as rung two however the tag names are unrelated to the logic. Who knows what the first rung is supposed to accomplish.

Believe it or not programmers do this all the time, take the following scenario as an example. Edward works the night shift at the bubble gum plant. The bubble gum expander machine locks up and won't expand the gum, a disaster. Edward, being the savvy programming type, opens up the program and adds an XIC instruction that fixes the problem. All good and well, but rather than defining a tag with a descriptive name he uses his own name.

To him this made sense, if anyone needed to see his awesome programming skills they could find it in the code. Everybody would surely be impressed by his witty contribution to the bubble gum expander machine. In reality Edward just made the code hard to read, more illegible. Don't be an Edward, create tag names that convey the purpose of the tag.

Structured Data Types

A logical group of tags is known as a structure. A User Defined Type or UDT for short contains as least one member. A member is simply a tag of any data type. A structure can have multiple members, each with its own unique data type.

Structured data types are easily identified as they have a characteristic "." separating a tag name from its members. Complimenting UDT's are pre-defined and module-defined data types. A UDT is defined by you the programmer, pre-defined and module-defined types come with the software and are available for use in instructions and modules. Take for instance the pre-defined data type TIMER. When a timer instruction is used the instruction will require an argument of the data type TIMER. *A pre-defined instruction will require a pre-defined data type.* The TIMER data type has all the necessary members for the timer instruction to operate.

Members: Data Type Size: 12 byte(s)

Name	Data Type	Style	Description	External Access
PRE	DINT	Decimal		Read/Write
ACC	DINT	Decimal		Read/Write
EN	BOOL	Decimal		Read/Write
TT	BOOL	Decimal		Read/Write
DN	BOOL	Decimal		Read/Write

When enabled, a timer instruction will update the accumulator (.ACC) argument. When the accumulator is greater than or equal to the preset (.PRE) the done argument (.DN) is set. The .EN and .TT arguments are Enable and Timer-Timing respectively and are updated when the timer is enabled and or timing.

User Defined Data Types

Take a look at the following tags, each tag is used to describe a recipe in a machine. The machine will behave differently depending on the recipe selected. For instance this theoretical machine assembles Hubble Bubbles. As everyone knows Hubble Bubbles come in three different colors, sizes and shapes. The nine tags below are used to describe the attributes of the three different Hubble Bubbles.

- Bubble_Bling_Color DINT
- Bubble_Bling_Shape STRING
- Bubble_Bling_Size REAL

- Bubble_Zing_Color DINT
- Bubble_Zing_Shape STRING
- Bubble_Zing_Size REAL

- Bubble_Waz_Color DINT
- Bubble_Waz_Shape STRING
- Bubble_Waz_Size REAL

The three different Hubble Bubbles are Bling, Zing, and Waz. Each Hubble Bubble has a color, shape and size attribute that will be used in the ladder program. The program will ensure that the correct parts are being used to assemble the correct Hubble Bubbles by utilizing the appropriate tags.

As an alternative to creating nine separate tags it makes more sense to create a UDT with all the relevant members. Remember a UDT is nothing more than a collection of data types with unique names. I created a data type with the name *Hubble_Bubble_Recipe* and populated it with the following members:

- Bubble_Color DINT
- Bubble_Shape STRING
- Bubble_Size REAL

Defining UDT's is fairly straightforward, once a UDT is defined it can be used to create tags of the data type Hubble_Bubble_Recipe, just as you create a tag of type DINT or TIMER. Instead of creating nine different tags I can create three different tags that contain all nine members. The three tags I create will be named Bing, Zing and Waz and be of the data type Hubble_Bubble_Recipe rather than something like BOOL or DIINT.

- Bing.Bubble_Color DINT
- Bing.Bubble_Shape STRING
- Bing.Bubble_Size REAL

- Zing.Bubble_Color DINT
- Zing.Bubble_Shape STRING
- Zing.Bubble_Size REAL

- Waz.Bubble_Color DINT
- Waz.Bubble_Shape STRING
- Waz.Bubble_Size REAL

Array Data Types

When you define a tag not only do you select how much memory the tag will consume by selecting a data type, you also select how the memory is arranged. Take for instance the user defined type Hubble_Bubble_Recipe. As an alternative to creating three different tags named Bing, Zing and Waz we could create one tag and define the tag as an array.

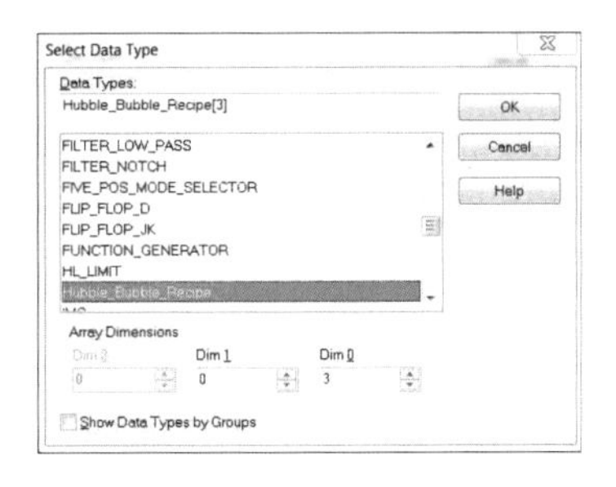

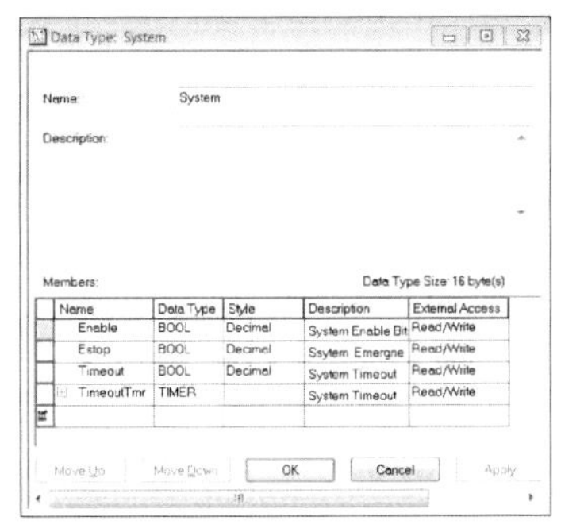

In the screenshots above I created a single dimension array with three elements defined. Arrays can have up to three dimensions. For now we will concentrate on single dimension arrays. I gave the following tag the name *Recipe* and defined it as an array of three of the type Hubble_Bubble_Recipe. Arrays are easy to spot as they contain the characteristic brackets [0]. The number inside the brackets is referred to as the subscript.

- Recipe [0].Bubble_Color DINT
- Recipe [0].Bubble_Shape STRING
- Recipe [0].Bubble_Size REAL

- Recipe [1].Bubble_Color DINT
- Recipe [1].Bubble_Shape STRING
- Recipe [1].Bubble_Size REAL

- Recipe [2].Bubble_Color DINT
- Recipe [2].Bubble_Shape STRING
- Recipe [2]Bubble_Size REAL

Defining the tag as an array creates a contiguous location in memory to store all the defined data. What does this mean? It means the bits occupy memory without any gaps in between. This comes in handy when you start using lookup tables and indirect addressing.

There is more to defining a tag then giving it a name. Creating a tag defines an address, how much memory is used, how that memory is organized, and how the data is accessed. A tag data type can be Atomic, array, structure, or a combination of all or some.

As a side note some of you will have noticed that when we defined the tag as an array we lost the descriptive names of the recipes, a perceptive observation. The descriptive name of the tag was replaced with a bracketed number or subscript. This is easily corrected by adding an additional member to the data type *Hubble_Bubble_Recipe* which will update the tag *Recipe*. By adding the new element Bubble_Name and defining it as a STRING I can effectively populate the tag with a descriptive name making my code more legible.

- Bubble_Name STRING
- Bubble_Color DINT
- Bubble_Shape STRING
- Bubble_Size REAL

The interesting thing about an array is how it can be addressed in code. The subscript allows elements of a tag to be addressed rather than the entire tag. This comes in handy in indirect addressing. For instance Recipe[0].Bubble_Color points to the same address as Recipe[Variable].Bubble_Color assuming that the tag *Variable* contains a value of 0. What if the tag *Variable* contained a value of 1 or 2? This type of addressing is what is referred to as indirect-addressing. The tag *Variable* is used to pick the subscript of the array.

Tag Scope

We will cover structures and arrays and the pros and cons of both more in depth as we progress through the book, for now I want to touch on scope, and not the kind you use for good oral hygiene.

Defining the scope of a tag is nothing more than defining its boundaries. There are two possible selections one can make when defining tag scope. Tags can be declared as controller scope or program scope. Tags defined as controller scope are visible to every program in the controller, there are no boundaries. Tags defined as program scope are only visible to the program it is defined in, hence the boundaries.

Remember a tag name is nothing more than an address to a memory location. That means that every tag must have a unique name in order to point to a unique memory location, right? This is true if the tag is defined as controller scope, however program scope tags can have the same name as tags defined in other programs local to a project. The tags can share the same name, however the actual address in memory is different due to the boundaries of the programs.

Important: A program scope tag and a controller scope tag can share the same name. The controller will always use a program scope tag over a controller scope tag of the same name. You can get yourself into a lot of trouble if your intent is to use a controller scope tag that shares a name with a program scope tag.

It is ok to give tags the same name only if the names are defined in different programs and the scope is set to program. It is also allowable to define a controller tag with the same name as a program scope tag, however don't do it, no good will come out of it. Doing this is confusing and can lead to the wrong tag being used in the wrong place.

Summary

As you can see there is a lot to variables. Variables are a core component in any programming language. After reading this chapter you should be able to identify common data types including Atomic, structured (UDT and Predefined) and arrays. You should also have an excellent grasp on the concepts of tag scope and memory organization.

Tags address memory. The tag definition process defines how much memory is allocated to a tag and how the memory is organized. Tags are referenced by instructions. Input instructions use the data addressed by tags as conditions. Output instructions change the data addressed by tags. Got all that? Good, then you're ready for a little Q&A.

Chapter 3 Questions

1. In the controls world variables are referred to as _______.

- Tags
- Instances
- Strings
- Parameters

2. When a tag is defined the name of the tag is used as an argument for _______.

- Defining classes
- Calling function blocks
- An instruction
- Polymorphism

3. True or false, when defining a tag a descriptive name is not that important, that is what the description is for.

4. How are Atomic or native data types best described?

- How many members are in the data type
- Data types that have a native number of members
- Data types that are defined as standard tags.
- The most primitive data types available including BOOL, SINT, INT, DINT and REAL.

5. True or false, structured data types may not contain other structured data types.

6. A structured data type in RSLogix 5000 includes.

- Pre-Defined tags
- User defined tags
- Module defined tags

- Add on defined tags
- All of the above

7. Which of the following tags is a member of a structure?

- MotorRun
- Motor_Run
- Leroy.Tag
- Recipe[1]
- None of the above

8. Which of the following best describes an array?

- A collection of tags
- A tag with a subscript
- A tag that consumes contiguous memory addressable by a subscript
- A pre-defined tag

9. True or false, when configured properly any program in a project can access a program scope tag.

Chapter 3 Answers

1. In the controls world variables are referred to as tags.

   Tags are an address to a location in memory.

2. When a tag is defined the name of the tag is used as an argument for an instruction.

   Tags and instructions are the building blocks of ladder logic.

3. True or false, when defining tags a descriptive name is not that important, that is what the description is for.

   False. A descriptive name makes the code more readable. Descriptions are not downloaded to the PLC they are stored on the programming computer. If you upload a project without an offline copy the descriptions will be lost.

4. How are Atomic or native data types best described?

   The most primitive data types available including BOOL, SINT, INT, DINT and REAL.

5. True or false, structured data types may not contain other structured data types?

   False, structures can be nested more than once, as a matter of fact.

6. A structured data type in RSLogix 5000 includes.

   All of the above, a structure is easily identified by the "." separating the members.

7. Which of the following tags is a member of a structure?

Leroy.Tag Leroy is the tag name and .Tag is one of its members.

8. Which of the following best describes an array?

   A tag that consumes contiguous memory addressable by a subscript.

9. True or false, when configured properly any program in a project can access a program scope tag.

   False. Program scope tags are visible only to the program where it is created in. Controller scope tags are visible to all programs in the controller.

C-4 Program Scan & Instructions.

Regardless of the application every program operates the same. In Ladder Logic a program is made up of a collection of instructions and variables placed in such a way as to create a logical program.

Ladder programs are scanned from top left to bottom right just as you are reading this book. When a page in a book is read you start from the upper left word and read through to the lower right word. This is done from the beginning of the book to the end. When the book is completed the reader will either be a little more informed or a little less inclined towards academics. For instance, if it was a juicy romance novel the reader may be a little less interested in programming and a lot more interested in pirate wenches; if it was a great book on programming, like this one, the reader will be much more intelligent and prone to donning smoking jackets whilst sipping cognac. Right? Of course I'm right.

What Does the Program Scan For?

When you read and write logic keep the flow in mind. Think of the program scan as flow from one instruction to another. The first instruction is scanned, then the next and so on until every instruction in the program is scanned. Of course this is not always true, for example there may be a conditional instruction that tells the scan to skip a block of code. Perhaps in the next scan the same instruction directs the processor to scan the previously skipped code. The execution of the program scan is therefore configurable using instructions. The code that is scanned is configurable, however the scan itself is not, it will always follow the top to bottom left to right rule.

Input & Output Instructions and Their Relation to Tags

All instructions are not created equal, for instance take the OTE and OTL instructions. Both are used quite extensively in ladder logic. The instructions write either a 1 or 0 to their associated arguments. Take the following example of an unconditional rung, there are no input conditions on the rung, hence the term unconditional rung.

When the rung is scanned the value of the tag OTE_ARGUMENT is unconditionally high. What happens to the argument when a condition is added to the rung?

The argument OTE_ARGUMENT is high when the argument XIC_CONDITIONAL is high. If the argument XIC_CONDITIONAL changes to low what happens to the value of the argument OTE_ARGUMENT? When the rung is scanned the XIC instruction examines the value of the argument XIC_CONDITIONAL which is low. The flow to the OTE instruction is removed, resulting in the OTE instruction changing its argument to false. This type of rung can be called a conditional output rung. The output follows the input. If the input or inputs are true then the output is true. If the input or inputs are false then the output is false.

What about the Output Latch (OTL) instruction. An OTL instruction behaves similar to an OTE instruction when placed unconditionally on a rung, the argument OTL_ARGUMENT is driven high.

When the same conditional XIC instruction is placed in front of the OTL instruction how will the OTL instruction behave?

When the value of the argument XIC_CONDITIONAL is low then the value of the argument OTL_ARGUMENT will be low, similar to the OTE instruction demonstrated above. When the XIC_CONDITIONAL argument value is high the OTL instruction will change the value of the argument OTL_ARGUMENT to high. What happens to the argument OTL_ARGUMENT when the value of the argument XIC_CONDITIONAL changes to low? The value will remain high unlike the argument of the OTE instruction. An OTL instruction is a latch instruction while an OTE instruction follows the rung input conditions.

Latch instructions and conditional instructions behave differently. When using an OTE instruction never reference the same tag with more than one instruction. Why, you ask? Look at the following rungs and decide for yourself. Can you identify the problem here?

Suppose Shauna is writing a program and needs the tag DIAL_INDEX to turn on when an index button is pushed or an AUTO_INDEX is detected. At first glance the rungs above look as though they will get the job done. In the first rung, when the button is pushed the argument DIAL_INDEX is driven high. On the second rung, when the argument AUTO_INDEX is true the argument DIAL_INDEX is driven high, however the logic above will not work, but why? Look at the rungs again.

The first rung evaluates as true which writes a 1 to the argument DIAL_INDEX. In the second rung the rung evaluates as false which writes a 0 to the argument DIAL_INDEX. In reality the argument DIAL_INDEX is switched on then immediately switched back off. The logic doesn't work. What if the second rung evaluated as true and the first evaluated as false? In this case the argument DIAL_INDEX would be switched off then immediately turned back on. It would appear to be on all the time. The first rung will never have the ability to write data to the argument DIAL_INDEX, because the second rung will immediately overwrite the data.

What would happen if the OTE instructions were replaced with OTL instructions?

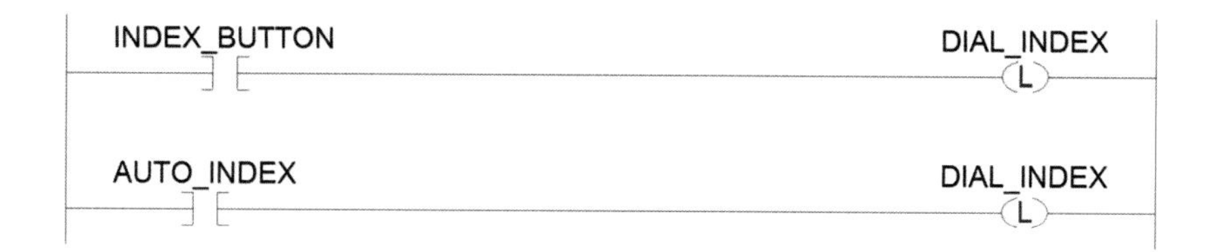

The logic will behave differently. When the INDEX_BUTTON is pressed the argument DIAL_INDEX is driven high by the OTL instruction. When the program scan reaches the second rung and the AUTO_INDEX is false the scan continues to the next rung in the program without changing the value of the argument DIAL_INDEX. The tag will remain true until it comes to an Output Unlatch (OTU) instruction that references the argument DIAL_INDEX. OTL arguments will remain high until driven low by an OTU instruction.

What if you don't want to use to use OTL and OTU instructions? Is there another way of writing the logic using just one OTE instruction? Of course there is, simply use a branch and create OR logic.

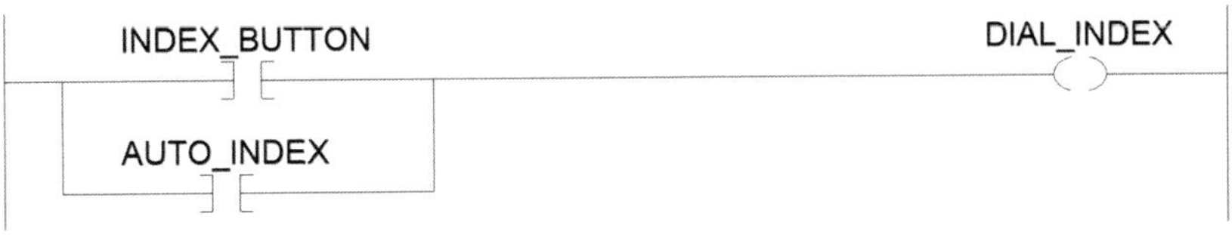

INDEX_BUTTON
DIAL_INDEX
AUTO_INDEX

If the INDEX_BUTTON or the AUTO_INDEX argument are true the argument DIAL_INDEX is driven high by the OTE instruction. When both input conditions are false the argument is driven low.

The same rule can be applied to other instructions as well. Take for instance the Timer on (TON) and Retentive Timer (RTO) instructions. The TON instruction is conditional while the RTO instruction is retentive. A TON instruction is reset when the conditional logic preceding it is false. An RTO instruction is reset with a Reset (RES) instruction.

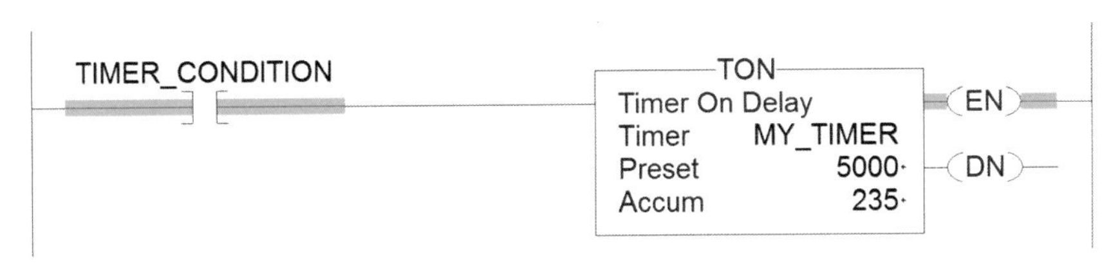

A TON instruction will write data to the argument MY_TIMER as long as the input rung condition remains true. When the rung condition is false the TON instruction is reset, the TON instruction resets the values of the argument MY_TIMER.

If the TON instruction is replaced with an RTO instruction the reset procedure is different. The RTO instruction is retentive. When the input condition goes false the RTO instruction does not reset the values of the argument MY_TIMER, the data remains unchanged. If the rung condition becomes true the instruction will again update the argument continuing the timer countdown.

Move & Copy Instructions

A move (MOV) instruction merely copies the value of one tag to another. The name of the instruction implies that it moves a value from one place to another, which is not entirely true. In reality the data is copied from the source tag to the destination tag. The value in the source tag remains unchanged.

So what then does the copy (COP) instruction do? Similar to the MOV instruction the COP instruction copies data from the source tag to the destination tag. The difference between the two is how the data is copied. You will remember that defining a tag defines the address, memory usage, and how the data is organized. As an example a structure can have multiple members of varying data types. The COP instruction can be used to copy all of the data in the tag from the source to the destination. The tags Bing and Zing are defined as the type Hubble_Bubble_Recipe and contain the following members:

- Bing.Bubble_Color DINT
- Bing.Bubble_Shape STRING
- Bing.Bubble_Size REAL

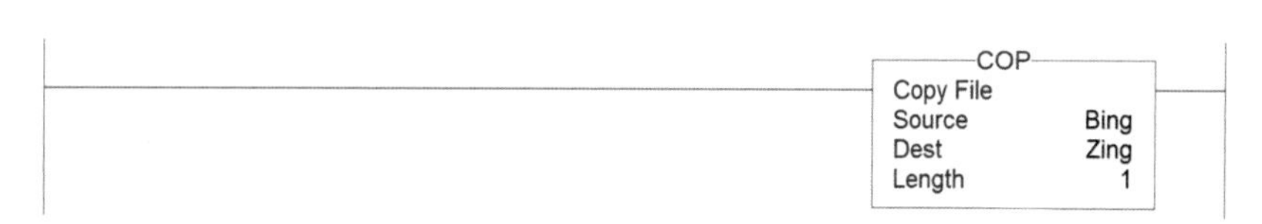

The above instruction will copy all the data from the tag Bing to the tag Zing, including all the members color, shape and size.

The COP instruction is used to copy a group of data also called a file. The COP instruction can copy complex tags such as UDT's and arrays. The length attribute defines how many elements to copy. The source tag and the destination tag members should be of the same data type, otherwise who knows what will happen.

The MOV instruction is used to copy Atomic data types; SINT, INT, DINT & REAL. To copy only the Color element from the tag Bing a MOV instruction is used.

Multiple COP and MOV instructions can be referenced to the same tag. It is ok to have a program with several MOV or COP instructions that reference the same argument.

Summary

Inputs such as the XIC and XIO instructions do not write data to an argument, therefore multiple input instructions can reference the same argument. Output instructions such as the OTE, OTL, TON, and RTO instructions write data to arguments. You must therefore understand how they work and use them wisely. Never address the same argument to more than one OTE or TON instruction, which is true for many output instructions. The OTE and TON instruction were used to illustrate this point. Some instructions like the OTL, MOV, & COP instructions can be used multiple times while referencing the same argument.

Chapter 4 Questions

1. True or false, Ladder logic is always scanned from left to right, however the top to bottom scan is configurable.
2. True or false, it is ok for multiple OTE instructions to reference the same tag.
3. True or false, it is ok for multiple OTL instructions to reference the same tag.
4. True or false, the same tag can be referenced by multiple input instructions.

Chapter 4 Answers

1. True or false, Ladder logic is always scanned from left to right, however the top to bottom scan is configurable.

   False, Ladder logic is scanned similar to how a book is read, from top to bottom, left to right.

2. True or false, it is ok for multiple OTE instructions to reference the same tag.

   False, a tag should be referenced to an OTE instruction only once. An OTE instruction is an output meaning it will change the data of the referenced tag. Using the same tag reference on multiple OTE instructions will result in unexpected results.

3. True or false, it is ok for multiple OTL instructions to reference the same tag.

   True, the same tag can be referenced to multiple OTL instructions. An OTL is an output instruction meaning it will change the data of the referenced tag, however it is a latch instruction so the data will be retained until it is cleared with an OTU instruction.

4. True or false, the same tag can be referenced by multiple input instructions.

   True, it is perfectly acceptable for more than one input instruction to reference the same tag.

C-5 Creating A New Project

"The problem with programs is they do exactly what you program them to do, not what you want them to do."

By now you should have an excellent grasp on the various components that make up code. You should be able to identify instructions, variables, and rungs. Now it's time to put these components together to create a project, but where to start?

Starting a New Project

When I first got into this business of writing code I knew how to call a subroutine, but I didn't know why, or when I should call it. When examining a program I always wonder why it was organized in a particular way. Was it merely preference, or was it required to make the project work? What about this whole task business; what kind of task should be used? How many programs and routines should a project have? Should parameters be passed to subroutines? Why? I had been reading about all these elaborate instructions and thought I should be using them in my code, however I didn't realize one thing... *I was complicating the process.*

Some of you will start with a project that has been commissioned, you may simply have to troubleshoot or modify it from time to time. Some of you will find yourselves adding new devices and code to existing projects. Still others will be removing and consolidating devices and code. Regardless of where you end up, a great place to start is by creating a simple project yourself, or at the very least get an over-the-shoulder view of the process.

Keeping It Simple (Within Reason)

The key to programming is K.I.S.S. or, Keep It Simple Stupid. This is what I will demonstrate in this section of the book, teaching you how to write code while keeping things simple. It adds no value for me to show you all my crazy programming skills. I could create a program using every available instruction, however I use only a handful of instructions with most programs I write. I could create an entire project with one routine and a half dozen rungs, however this would be contrary to the one rule that keeps me out of trouble. If you can make a rung so complicated that not a single one of your peers can interpret it, congratulations you just created an example of what not to do. Making legible code is just as important as making code that works. Write code in such a manner that any programmer can read it.

Have you ever gone on a road trip to somewhere you have never been? When I go on a trip I plan ahead, I use a map that allows me to plan my route. I can estimate time, distance and plan for some sightseeing along the way. Writing code really is no different. If you know your origin and your destination the only thing left to do is decide how you will get there.

You need to plan your route. What will the code ultimately do, what is its purpose? Is it going to move freight, package widgets, assemble parts, load pallets? Answer this question and you have your destination.

Is this a new project, a retrofit, or an addition to an existing project? This is your starting point, your origin. Once you have these two questions answered you have the base of your map, all that's left to do is plan your route. The code will solve a problem, it lies between the origin and the destination.

A Freight System Description

In this book we will explore building logic that will operate a conveyor system. The system will be fairly basic, freight is placed on an input conveyor and transferred to the opposite end of a warehouse where the freight is queued and manually removed.

Creating logic for this fictitious system will help you understand how a project is put together.

The system will be comprised of six conveyors each with a constant rate of 100 feet per minute. The longest package to be conveyed will be 18 inches. The first conveyor in the system will be the load conveyor. The next four conveyors will transfer freight to the offload conveyor at the end of the system. Sounds simple enough, right? Each conveyor will be outfitted with a motor, Motor Service Disconnect, Emergency Stop and discharge photoelectric sensors. The sensors will be used to turn conveyors off and on depending on downstream conditions and provide Jam protection. The system will be started from a control station located at the load conveyor. The control station will have a momentary key switch that will enable and start the system. The system will remain enabled until all freight is unloaded, at which time the system will disable automatically.

C-6 Defining & Creating Logical Code

To get things started we will work out the logic visualizing what is needed to make the process work. This process takes trial and error to get it right. Once you get in the groove it goes fairly easily, it can be a bit of an art, the trick is to keep it simple.

The conveyor functionality is relatively simple. Once enabled a conveyor will turn on and off depending on the states of conditional inputs. This is a simple explanation of the functionality, however it can be broken down further. In order to transfer freight from one end of the conveyor to the other the conveyor must run.

Thinking in terms of logic, what is the simplest way to do this? What about an unconditional rung with an OTE instruction addressed to the conveyor motor? When the argument is driven high by the OTE instruction the conveyor will turn on. This is the simplest logic I can come up with to turn the conveyor on, and it was uncomplicated and easy.

During the logic building process I rarely define tags, rather I place an undefined name as a reference to what the eventual tag will be named. Don't worry about defining tags during this phase, it slows down the process of creating the all-important logic. Think of this phase as creating a rough draft of the final code. At this early stage defining tags make little sense as they will probably change names later on.

Now that the conveyor is up and running the next thing to do is define all the conditions that will stop the motor, this is the hard part.

If the system is disabled it will not run. The system is enabled via a momentary key switch and disabled automatically after all the freight has been removed from the conveyors. A conveyor will also stop if there is an active emergency stop, motor service disconnect, VFD fault, Jam fault, or Backup condition.

- System Off
- System Timeout
- Emergency Stop
- Motor Service Disconnect
- VFD Fault
- Jam Fault
- Backup condition (downstream conveyor is off and discharge PE is blocked)

Generate All the Input Conditions

Creating logic to make the conveyor run was easy, the trick then is to get the conveyor to stop when you want it to. The conveyor functionality can be defined with the following statements: There are seven distinct conditions that can stop the conveyor. If all conditions are met the conveyor will run, however if any single condition is not met the conveyor will stop. This sounds a lot like *AND* logic, if this and this and this are met run the conveyor, if not stop it. The following rung has all the conditions placed in series effectively creating an *AND* logical circuit. If all the conditions are met the output of the rung is true and if any conditions are not met the output is false.

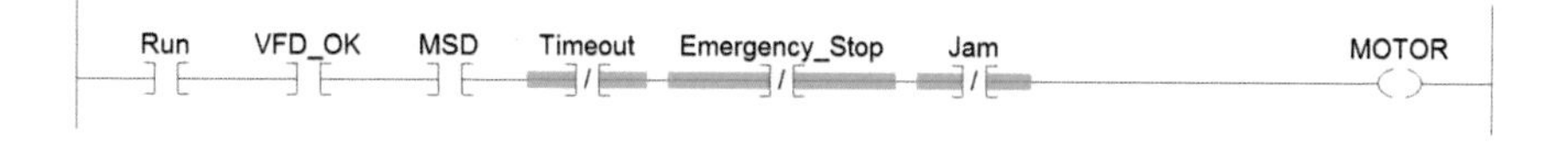

Group Like-Input Conditions

Now that the conditional *AND* logic is in place to control the conveyor it's time to better define the different stop conditions.

For instance the system timeout condition and the system off condition can be grouped together to create one condition because they really are two parts of the same thing. To get an idea of where I am going with this take a look at the following rung.

This rung logic states that if the timeout condition is not met (it has not timed out) then turn on the run output. The run argument will be used as a condition for the conveyor circuit in place of both the system off and system timeout conditions. This makes sense as the conveyors will be disabled when the system times out. This will satisfy the definition of enabling and disabling of the system. No conveyor can run unless enabled, and all conveyors will be disabled when all freight is removed.

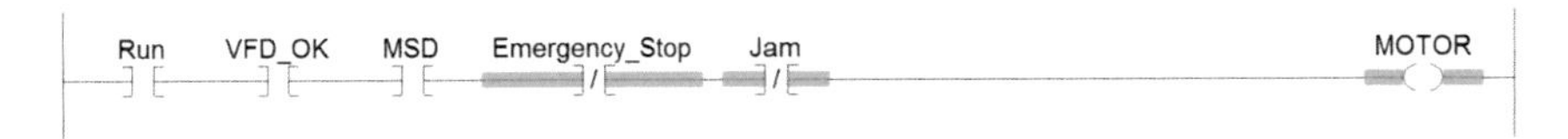

I know what you're thinking, "Big deal, all you did was move the timeout condition to a different rung, and you didn't really do anything significant here." Nevertheless there are several conveyors that will use the run argument, so in essence this made the conveyor logic easier to read, it made the final code smaller and more portable. Stick with me I'm going somewhere with this, I promise.

Notice that I could have placed other conditions in this group, but I didn't. For instance I didn't put the emergency stop in this group because I don't want to disable the system in response to an operator activating an emergency stop. An emergency stop is a temporary disabling of the conveyors, a condition that can be reset when the condition is removed. According to our specifications the system is enabled from a control station on the load conveyor and disabled when all conveyors have been cleared of freight. The only condition that will disable *all* of the conveyors is a system timeout.

In order to complete the system run logic the *run* circuit needs to stay true until all freight has been removed from the system. This is not the case for all conditions, for instance a Jam fault should not disable the system and applies to only a single conveyor.

The *run* circuit is nearly complete, it states that the system will stay running as long as there is no timeout, but how will the system be started? The specification states that the system can only be started with a momentary key switch similar to the key switch in a car. When the key switch is turned to the start position a starter motor is engaged to start the car, when the key switch is released the spring mechanism returns the key switch to the run position. Rather than engaging a starter-motor, the conveyor key switch will provide a discrete input to the PLC resulting in the enabling and subsequent startup of the system.

Adding the key switch condition will enable the system as long as the key switch is made, however as soon as the key switch is released the run argument will turn off. This is where a latch-in circuit will come in handy. This is also a great example of *OR* logic. The key switch condition, *OR* the run condition will keep the run argument high until the timeout condition transitions to true thus unlatching the circuit.

The latch-in circuit is a very common circuit and arguably one of the most important circuits you can learn because of its wide use and flexible functionality. As stated earlier it is perfectly acceptable to use the same tag as an argument for multiple input instructions. Using the same tag for multiple output instructions is a different story.

Define the Problem and Generate Simple Code

Now that the system run condition has been fully defined it is time to define the timeout condition. There are as many ways of coding this as there are programmers in the world. The logic controlling the timeout circuit could be made extremely complicated or relatively simple. Bear in mind that the bulk of your effort should not be spent in an attempt to overly simplify the code. Keep it as simple, easy and readable as possible. The beauty about freight systems like this one is that it is fairly easy to determine if freight is still present on the conveyors. "No freight left behind", that is my moto. Let's examine a couple of different scenarios that we could use to determine when the system is clear of all the freight.

In the first scenario, logic could be written to keep a count of the freight in transit using a combination of Count Up and Count Down instructions (CTU & CTD). Utilizing these instructions the logic could keep a count of the freight that is in transit. A count could be incremented when freight is put on, then decremented when freight is delivered, however this can be unreliable. For instance, what if there is a Jam condition and someone manually removes a couple of packages from the system before the removal point where the logic decrements the count? The system would never detect that the freight was removed and keep running, possibly forever. This same thing would happen if we were to use a FIFO (First in First Out) system as well.

In either case we could add some Band-Aid code that goes something like "Hey I think this package should have been here by now, but it hasn't made it yet so just decrement the count by one." Band-Aid code would work, but I think our freight system deserves better, more robust logic.

Another scenario might be to positively track every package on the system to its final destination. The system would know where all packages are at all times. If a package does not turn up when it should, the logic could decrement the count or stop everything until the package is located. This could work, however the project just got a lot more complicated. We would need more hardware; encoders, sensors, maybe even bar code readers. What about all the code that would need to be written, tested and debugged? This is definitely not the right path for this project.

Each conveyor has a discharge sensor that is triggered when a package reaches it. If there are no packages on a conveyor while the conveyor is running then the conveyor must be empty. These fictitious conveyors are fairly small, which means freight will be conveyed from one end of the conveyor to the other in less than a minute.

Put another way if a conveyor runs empty for at least a minute then it is most definitely empty. Just to be sure our code will allow for two minutes of empty run time to ensure the conveyor is empty and the individual loading freight on the conveyor is done.

We have effectively defined the timeout logic, now it's time to transfer the definition into logic. In order to effectively determine a conveyor is empty a conveyor must be running and the associated discharge photoelectric sensor must be clear for at least two minutes.

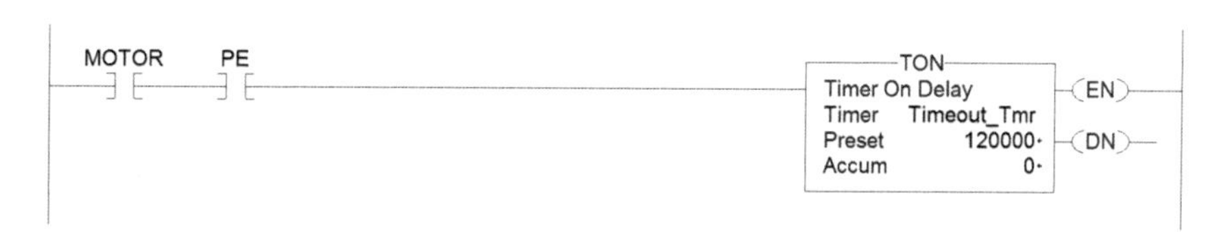

A Timer On instruction will accumulate milliseconds until the accumulator reaches the preset value. At this time the timers done argument is set. The TON instruction will reset its argument when the rung conditions are false. If a conveyor turns off or a sensor is blocked the timer is reset. The completed rung has a branch circuit where the timeout argument is set.

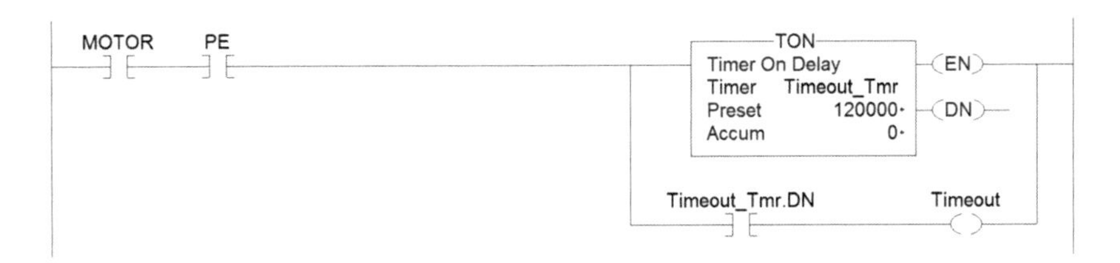

Compare the completed rung to the one before it. Either rung is perfectly acceptable, the latter rung is more readable because it has a descriptive timeout argument, whereas the timer done bit can be confusing when used elsewhere in logic. The two rungs accomplish the same thing, however I would go with the second rung myself only because it makes the logic easier to read. For instance the Timeout_Tmr.DN argument could be tied to a timer or a counter. If it is tied to a timer it could be a TON or a TOF which could yield different results depending on the instruction. The done argument is Boolean so it is on or off similar to the Timeout_Tmr.DN argument, however using it elsewhere in the logic can sometimes lead to more questions than answers.

Below is the completed logic that will be used to turn the system run bit on and off. Two rungs control the enabling and disabling of all the conveyors. Simple legible code. I would like to point out that the system timeout circuit will eventually contain all the conveyor motors and photo sensors. I will put in all the conditions when the tag names are fully defined, right now we are just working out the logic.

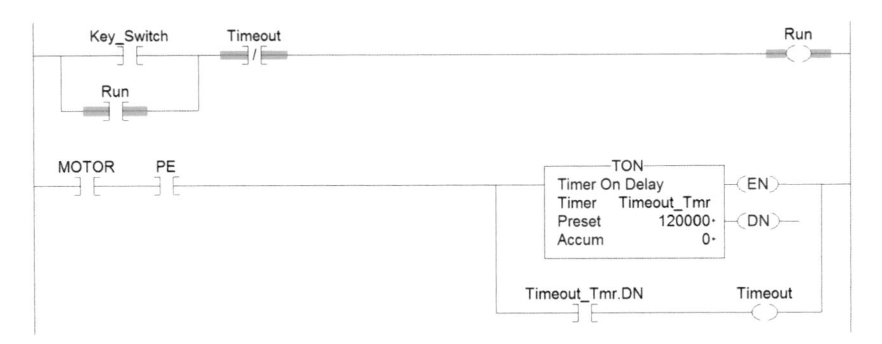

The OTL and OTU Instructions; Use Them Wisely

The next condition to consider is the Emergency Stop. The emergency stop condition will stop all conveyors in the system when activated. This condition will not enable or disable the conveyors like the *run* condition. Other conditions like a Jam, MSD or Backup condition will stop each conveyor individually.

The conveyor logic with respect to the Emergency Stop condition can be interpreted as follows; run the conveyor as long as there is no emergency stop condition. That part is easy and already done, but how will the Emergency Stop circuit work? An Emergency Stop condition is activated by an operator depressing a maintained Emergency Stop push button. The condition is reset when the Emergency Stop push button is pulled out and a momentary Restart push button is depressed.

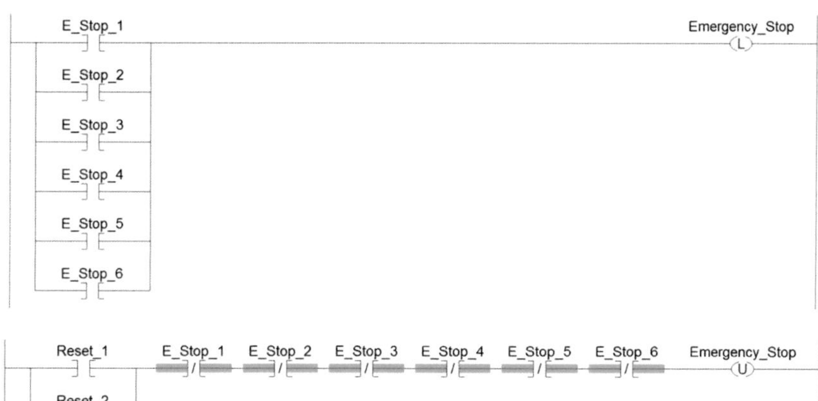

Can you identify what is happening in the rungs above? The first rung begins with OR logic. Emergency Stop 1 or 2 or 3... will activate an emergency stop. The second part of the rung is the output instruction OTL. The latch argument will stay active even when the input conditions become false.

The second rung will unlatch the emergency stop condition when a restart button is depressed *AND* there is no active emergency stop.

Why use the latch and unlatch bits, couldn't this logic be written similar to the *run* logic created earlier? The answer to the question is yes, however I created the emergency stop circuit this way for two reasons. The main difference between the two circuits is that the OTL instruction will maintain its last condition on power up. The latch-in circuit will not retain its last condition on a power up, it is reset.

Imagine the conveyor system is enabled, it is running and transferring freight. Lightning strikes a substation consequently knocking out power to the system. It's Friday at 2:00 pm and according to the utility company the power will not be restored until Saturday late morning at the earliest. The conveyor system is manually cleared of product and all employees are sent home. No big deal, disaster adverted.

The run circuit was programmed with a latch-in circuit rather than an OTL instruction. When the power turns on the logic is reset and the conveyor system is disabled. If the run circuit would have been programmed with the OTL instruction the conveyor system would have been enabled when the power was restored. The system would automatically be enabled, a big no-no.

Take the same scenario with an emergency stop condition. Suppose that the power outage occurs while the system is in an emergency stop condition. An operator depressed an emergency stop button and pulled it out in anticipation of restarting the system. The power goes out, everyone is distracted, and the system is cleared of product. The power comes back on and the emergency stop latch is still activated. In other words this condition is not reset by a power down condition, exactly what we want here.

On the surface both circuits appear the same, however digging into them you can see a very distinct difference. Be careful with latch bits. If used improperly, equipment could start when you don't want it to. A power outage should not reset a circuit intended for safety. In most cases after a power-up condition systems that control motion should never automatically start. Imagine if the conveyor motor outputs were OTL instructions rather than OTE instructions. If a conveyor was running when the power went out that same conveyor would start running as soon as the power was restored, no good.

It's also important to note that safety circuits such as an Emergency Stop should be programmed properly, however in respect to safety it's all about the hardware. Safety switches, relays, circuits and other appropriate hardware should be used in the design, maintenance and operation of safety circuits. Safety circuits should be designed by qualified electrical engineers. The above hypothetical scenarios are used as teaching examples not as an example of how to design a robust safety circuit.

XIC and XIO Instructions, Why *NOT*

Back to making the conveyor *NOT* run. Thus far we have done a good job simplifying within reason the logic that stops a conveyor from running. Using this technique we have also created some of the overhead logic needed to make the system work.

We have fully defined two conditions that will stop the conveyor from running a *Run* and an *Emergency Stop*. One condition enables all conveyors and the other inhibits the running of the conveyors. The *Run* condition is a compilation of a timeout circuit and a start circuit.

The next thing to think about are system faults. From the list made earlier we know an MSD, Jam and VFD fault should stop the conveyor.

Run Emergency_Stop VFD_OK MSD MOTOR

Both the MSD healthy and VFD healthy conditions are placed directly on the conveyor run circuit. Notice that the instructions referencing the conditions are both XIC instructions. If the MSD and VFD are healthy run the conveyor. Why did I elect to use a high healthy signal rather than a high fault signal? Simple if the VFD or MSD circuits fail how will they send a fault signal? If they fail they cannot send a healthy signal and the logic will not try to start the conveyor.

The Motor Service Disconnect is provided so that the conveyor motor can be disconnected from the power source for maintenance, removal and repair. A typical industrial conveyor motor receives power from a three-phase power source. The Motor Service Disconnect allows a service technician to disconnect power and lock out the power source. An auxiliary contact utilizing control voltage is typically wired up to an input on the PLC for use as feedback. This auxiliary contact lets the PLC know if the disconnect is on or off. The logic would expect to see the MSD condition high when the MSD is on and low when it is out of service.

Using a high signal as opposed to a low signal communicates to the PLC that the control voltage is operational and the MSD is in service (switched on). The MSD could be wired the opposite way. The auxiliary input would be high when the MSD is off and low when on. If this were the case what would happen in the logic if the control voltage to the circuit were to fail? The MSD signal would be off, thus the logic would attempt to run the conveyor even though the control voltage is missing. Wiring correctly is referred to as a fail-safe condition.

As another example consider a Variable Frequency Drive or VFD. A VFD is nothing more than a high tech motor controller. VFD's can be programmed to output several signals that can be used as feedback in your ladder logic programs. Take for instance a VFD healthy signal. The signal would be high when the VFD is healthy and low when there was something wrong with the VFD. What if the signal was backwards? That is, the signal was high when there was a VFD fault. What if power was removed to the VFD, how would you ever get the fault signal? The ladder logic would see that there is no fault condition and try to make the system run.

Make sure your signals are the correct state for what you are doing. Your system will only be as accurate as you make it. Unlike programming apps for phones or tablets you have control of the electrical system as well. It's your job to make a system solid, both electrically and logically.

Robust Code, Because Things Will Go Wrong

The next condition on the list of definitions is the Jam condition. This type of protection is implemented on typical conveyor systems. Freight will "jam" up, typically at transition points. This can be damaging, both to freight and equipment. A sensor is mounted at the end of each conveyor. The sensor has two purposes, one of which is to detect a Jam condition and the other is traffic control.

The process for detecting a Jam is fairly simple. If the sensor detects freight while the conveyor is running for a predetermined time a Jam is declared. This is easy enough to figure out if you know the speed of the conveyor and the maximum length of the freight being conveyed. According to the specifications the maximum length of a freight container is 18 inches. The conveyor speeds are a constant 100 feet per minute which equates to 1200 inches per minute. Dividing 1200 by 60 (seconds) gives us 20 inches per second.

Now that we know all the conditions and have all the parameters for declaring a Jam condition the only thing left to do is write it down. Writing down the problem in its simplest form is an excellent place to start forming logic. From there it's a matter of transferring it from one language to another, English to Ladder Logic.

If the sensor is blocked and the conveyor is running for more than a second declare a Jam.

The statement above can be simplified even more using just the key words.

Sensor blocked, AND Conveyor running, Time, declare Jam.

With the simplified statement it should be obvious what the logic will look like. The keywords *sensor blocked* and *conveyor running* are the two conditions that need to be satisfied to declare a Jam. The conditions are Basic *AND* logic. The PE signal is low when blocked and high when not blocked, another example of fail-safe logic. Think about it what would happen if the signal was the opposite way and the sensor was damaged?

PE MOTOR

Using the remaining key words *timer* and *declare Jam*, the remainder of the rung can be completed with a timer and a Jam declaration.

PE MOTOR
TON
Timer On Delay
Timer Jam_Tmr
Preset 2000
Accum 0
EN
DN
Jam_Tmr.DN
Jam

The above rung is not the only way to write the logic. In many cases this same logic will look like the rung below.

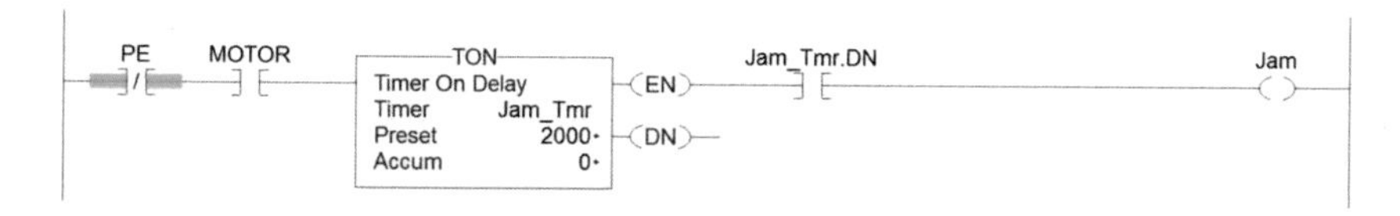

Both rungs perform identically, neither is the right or wrong way of constructing the code. The rung with the branch is backward compatible with previous versions of the language, other than that it's all about preference. I typically program using the branch method rather than stacking the logic out of habit.

Now that the logic to declare a Jam is in place how will it be reset? Obviously the Jam should be cleared and the freight will need to be manually removed from the system. Imagine pressing the restart button with the Jam condition still in place. Having logic that simply resets a Jam condition without first checking to ensure the Jam is clear would be a recipe for disaster. The Jam logic needs to be intelligent enough to prevent the conveyors from starting when the condition still exists. Once the obstruction is clear the Jam condition should be reset with a restart button.

Again, simplify the definition and use it to create logic. According to the definition above the Jam condition should be latched. The current Jam logic does not latch, in other words once the obstruction is removed the condition is cleared. This is bad, the conveyor would start after the obstruction is cleared. The Jam should be reset only when the condition is clear and a restart button is pressed. Another perfect opportunity to use a latch-in circuit.

The Jam circuit needs to be modified to accommodate the latch-in circuit. Rather than trying to create the entire circuit with one rung I have opted to break it into two rungs. The Jam declaration will be moved to a separate rung that can be used for the latch-in circuit. Rather than declare a Jam when the timer is done on one rung the Jam will be latched on a separate rung. Take a look at the two rungs of logic that will define the Jam logic.

If the conveyor is running and the sensor is blocked for the preset time the timer will finish setting the .DN argument. In the next rung the .DN argument will turn on the Jam bit latching in the condition. So far so good, all that's left to do is put in the conditions necessary to clear the latch-in circuit. To clear the condition the sensor must be clear and a restart button must be pushed.

Simplifying the above statement results in the following keywords: *Sensor*, *AND*, *restart*.

Now let's walk through the logic assuming we have a Jam condition. We have all the conditions necessary to clear the latch. There is just one problem, can you see it? We have created *NOR* logic for clearing the latch when we should be using *AND* logic. The sensor must be clear *AND* the restart button must be pressed to clear the Jam condition. The rung above clears the condition if the sensor is cleared *OR* the restart button is pressed.

In previous examples instructions were placed next to each other in series to create *AND* logic, however here it does not work here. Why?

It's all about the type of instructions and the placement of said instructions. In this rung our two conditions are XIO instructions which can also be referred to as NOT instructions, as in not true.

To simplify this look at the following rungs. Do you see *AND* logic or do you see *OR* logic?

The first rung is *NOR* logic and the second *NAND* logic. Replace the XIO instructions with XIC instructions and the logic is changed. The first rung will be *AND* logic and the second will be *OR* logic.

Take a look at the following truth tables. The first two numbers represent two inputs and the number after the equal sign is the output of the two inputs. So, how do you know if it's *AND* logic or *NAND* logic? *OR* logic or *NOR* logic? Simple, look at the instructions that are being used. If there are two not instructions (XIO) in series then you're dealing with *NAND* logic, if there are two XIC instructions in series you have *AND* logic. The same goes with the *OR* and *NOR* logic.

- **And Truth Table**
- 1 & 1 = 1
- 1 & 0 = 0
- 0 & 1 = 0
- 0 & 0 = 0

- **Or Truth Table**
- 1 & 1 = 1
- 1 & 0 = 1
- 0 & 1 = 1
- 0 & 0 = 0

- **Nor Truth Table**
- 0 & 0 = 1
- 0 & 1 = 0

- 1 & 0 = 0
- 1 & 1 = 0

- **Nand Truth Table**
- 0 & 0 = 1
- 0 & 1 = 1
- 1 & 0 = 1
- 1 & 1 = 0

Back to our problem. Logically speaking both the sensor and the restart button are needed to clear the latch. This can be accomplished a couple of different ways.

In this demonstration the Jam argument will stay on as long as the timer is done. The Jam timer is reset as soon as the Jam is declared. The Jam argument will turn off the conveyor motor. The conveyor motor argument is one of the conditions of the Jam timer circuit. Once the Jam argument is set by the timer done argument it is latched in until both the sensor is clear *and* the reset button is pushed. The placement of instructions when programming is everything. To reset the Jam the sensor must not be blocked and the reset button must not be off (it must be pressed).

An alternative to the Jam reset circuit could be made by using the basic latch circuit and adding another rung and that solves the reset logic. Either circuit will work, it all boils down to what is more readable. Maybe an argument can be made that the following is easier to understand and I might agree with that, however as a programmer/technician you will likely come across multiple scenarios. Neither circuit is wrong, a lot of the programming you will see is merely a programmer's preference.

Traffic Control

I think this is a good time to point out that thus far we have created useful logic circuits using just five instructions. The XIC, XIO, OTL, OTU and TON instructions. Most programs use only a handful of instructions and nearly all programs will use these five.

There are specialty instructions for everything including controlling motion, crunching numbers, communications, string manipulations, the list goes on. Master the fundamentals first. Learn and understand the basic instructions, learn how they work and how placement determines the behavior or outcome of the logic, then you can begin utilizing multiple instructions.

Back to our conveyor logic, which is nearly complete. There is just one more condition left to add to stop the conveyor motor. The last condition states that the conveyor motor will turn on as long as there is *NOT* a Backup condition.

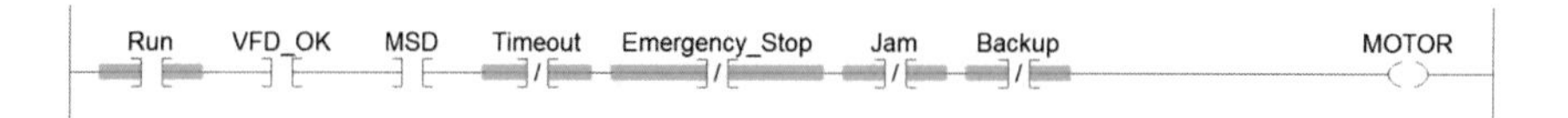

Following the basic rule for writing logic, the first thing to do is define what a Backup condition is. A conveyor must stop when freight reaches the end of the conveyor and the downstream conveyor is not running. This could happen if the downstream conveyor has a fault and is off. This condition is detected using a sensor mounted at the discharge end of the conveyor. This is the same sensor used to detect a Jam condition.

Now that the Backup condition has been defined it's time to simplify it by eliminating everything but the keywords: *Downstream NOT running, sensor blocked (not clear).*

The next step in the logic creation process is to take the keywords and turn them into instructions. *The sensor NOT clear* and the *downstream NOT running* tells us we have two NOT conditions. Another way to state it would be that we need two XIO instructions in series.

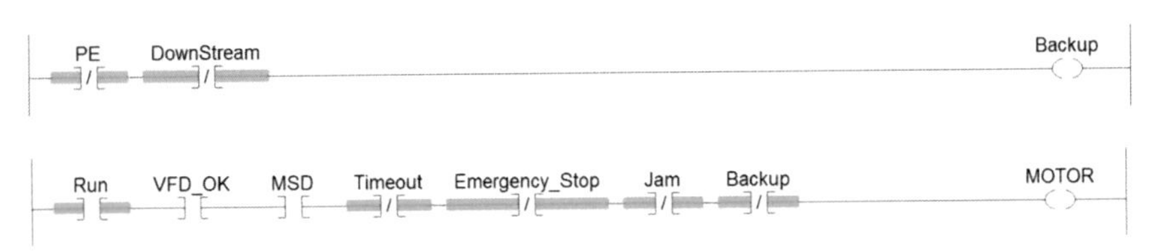

The Backup logic is nearly complete, but I think we can make this a bit stronger by turning the Backup logic into a latch-in circuit, but why?

There may be an instance when a small package will block the sensor and the downstream conveyor is off. What will happen if the package rolls past the sensor so it is no longer detected? In this scenario the Backup condition would be cleared allowing the conveyor to turn back on, thus pushing freight onto a conveyor that is not running. A very real problem with a simple solution. Once activated the Backup latch should stay on until the downstream conveyor starts running.

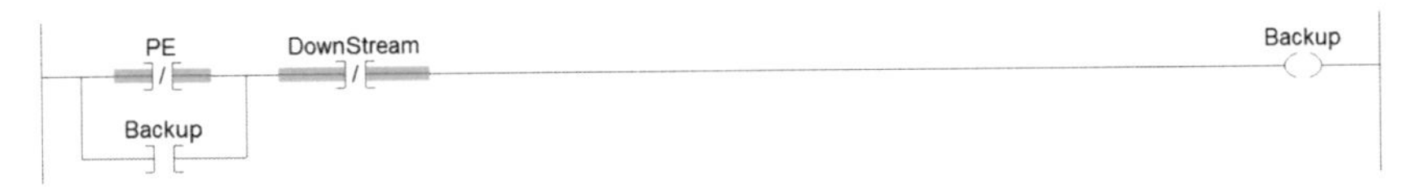

By adding the XIC instruction referencing the Backup we now have a Backup latch circuit. This will effectively keep the conveyor from running until the downstream conveyor starts back up, nice and controlled.

The sensor signal is high when no freight is detected and low when freight is detected, why not the other way around? The sensor signal would be high when freight is detected and low when no freight is present. What would happen if the senor was broken or removed? The logic would see that there is no freight there and just keep running. Again signal setup is everything, the sensors are set up for a fail-safe condition. If a sensor is broken or removed the system will declare a Jam and/or Backup condition, stopping the conveyor rather than just running it until all the freight piles up.

Wrap It All Up

Now that we have all our conditions necessary to stop the conveyor we need to think about a startup delay. Rather than starting the conveyor immediately it would be nice to first sound a horn and light to warn operators that the conveyor is about to start. This startup alarm should sound when the system is started or restarted.

To finish off the conveyor run logic we need to add a timer to the circuit to delay the motor from starting for five seconds after all the input conditions are satisfied.

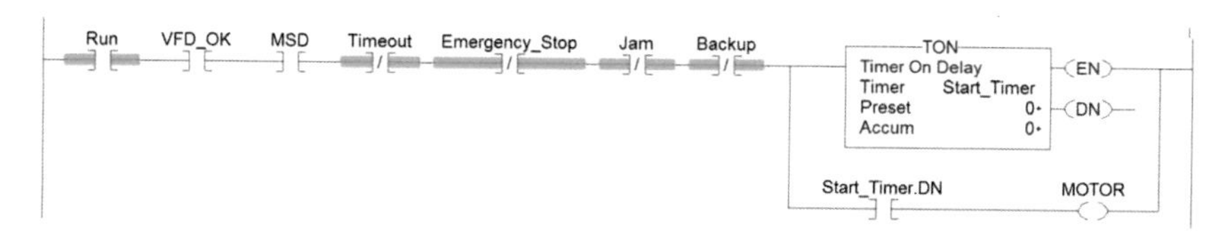

Using the timer done argument solves part of the problem. It effectively delays the start of the conveyor for five seconds after all input conditions are met. The other part of the problem is to sound an alarm when a conveyor is starting which can be done using the *timer timing* argument (.tt). The .tt argument is high while the timer is timing and off otherwise. An easy way to sound a horn and flash a light would be to *OR* all the .tt arguments with XIC instructions.

The completed rung looks great, however upon further investigation I realized one minor flaw. All the conditions with the exception of the Backup condition should trigger the startup timer and sound the alarm. The Backup condition is the only one that turns on and off automatically. It's more of a traffic control condition then a fault. When the Backup condition is active the conveyor will stop, however when the condition is removed the conveyor should start without the five second delay, alarms and horn.

Take a look at the conveyor logic rung, there is a very simple solution for this problem and I will give you a hint: "It is all about placement." Define the problem, examine the logic that is there, look at the key words and find the solution before moving on.

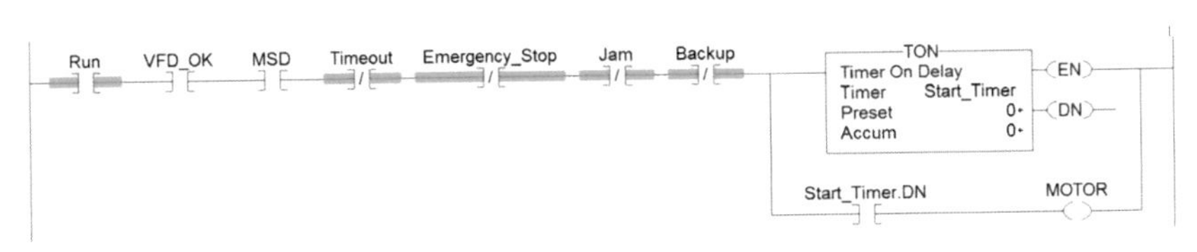

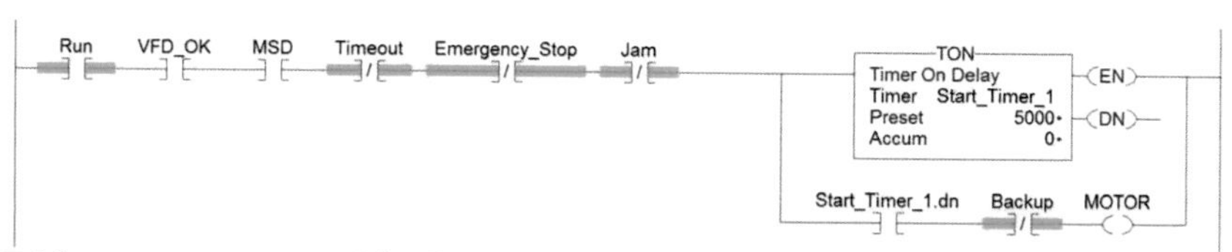

Did you come up with the same rung as me? I merely moved the Backup condition from the *AND* logic at the beginning of the rung and placed it after the timer done condition, but before the OTE instruction. Remember that a TON instruction is reset when the conditions before the timer are false. Moving the Backup bit still stops the conveyor, but does so in a manner that does not reset the timer.

Authoring and troubleshooting ladder logic can be frustrating if you don't understand what is going on, however once you understand that placement is just as important as the instructions it makes sense.

Summary

Thus far we have created a rung of logic that we will be used for every conveyor in the system. The *sys run* rung will be duplicated once for every conveyor. The logic is solid, however the arguments have yet to be defined. By creating the various stop conditions we created additional logic that will enable and disable the system. We also created logic for emergency stops, conveyor Jams, and Backup conditions. This is an excellent time to start thinking about how we should group the code with an emphasis on readability and portability.

Chapter 6 Questions

1. What comes first, defining tags or creating logic?

2. A TON instruction will accumulate __________ until the accumulator is greater than or equal to the __________.

- Quickly, system variable $Time
- Time, Accumulator
- Milliseconds, Preset
- Slowly, Sum of the Accumulator and Preset

3. True or false, an OTE instructions will be reset on a power up.

4. Simplify the following statements and create logical rungs complete with instructions and arguments.

- If the cup height sensor is on and the weld unit is ready to weld then run the rotation motor until the weld unit is complete.

- If the dead nest is empty run the vibe track and vibe bowl to bring a part to the dead nest.

- When the robot sends a request part signal to the PLC and a part is present at the dead nest send a part ready to pick signal to the robot.

Chapter 6 Answers

1. What comes first, defining tags or creating logic?

Create the logic first before defining the tag names.

2. A TON instruction will accumulate milliseconds until the ***accumulator*** is greater than or equal to the ***preset***.

3. True or false, an OTE instructions will be reset on a power up.

   True, the OTE instruction is reset on a power up condition. The OTL instruction will retain its last condition.

4. Simplify the following statements and create logical rungs complete with instructions and arguments.

   If the cup height sensor is on and the weld unit is ready to weld then run the rotation motor until the weld unit is complete.

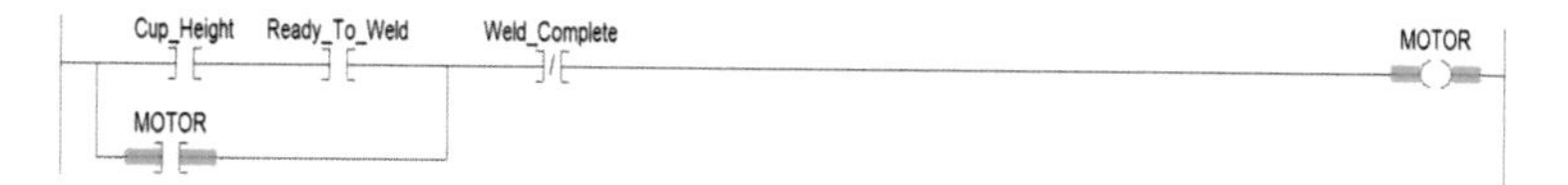

   If the dead nest is empty run the vibe track and vibe bowl to bring a part to the dead nest.

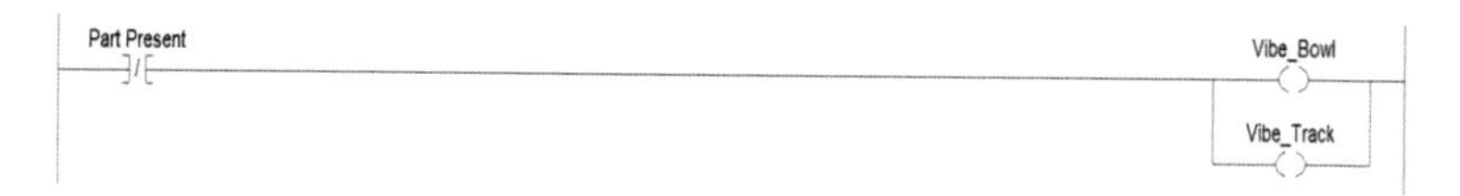

   When the robot sends a request part signal to the PLC and a part is present at the dead nest send a part ready to pick signal to the robot.

C-7 Tasks, Programs & Routines

Before completing our fictitious freight handling system and while the logic creation process is still fresh in our minds I think this is an excellent time to think about how to organize the code. How many tasks, programs and routines will our project have? Where will the various rungs reside in our program? Should they be placed in one routine, two routines or several routines; what makes sense?

Ladder Logic Programs

A program can be thought of as a container, one that houses all the local tags routines instructions and rungs relevant to the program. A controller project will have at least one program, however it can house many depending on the size and model of the processor.

In many cases one program is all that is needed to create a portable and legible project. In other cases it may make sense to create a project that contains several programs. As an example a number of years ago I worked on a freight handling system that controlled a number of conveyors as well as some peripheral equipment. One peripheral sorted the freight by transferring it vertically from one line to another, the other transferred freight to a parallel line. The interesting thing about this equipment is that both devices came with programs which I inserted into my project. All I had to do was tie these programs into my logic.

Rather than creating new code to control these peripheral devices I opted to use portable code developed by the OEM, which worked out well. Not only did I use standard code, I used it in a manner that allowed me to keep peripheral devices separate from my standard code.

A program can then be thought of as a library file that controls particular devices or processes. Imagine having a catalog of proven programs that control a multitude of devices; creating a project would be much easier than, say, writing every program from scratch.

Project Tasks

Programs are called by tasks. A program is executed either continuously, periodically, or from an event. This is configured by choosing the task that will govern the program.

- By Default programs are run continuously. They are called by the Continuous Task and when complete they will run again.
- A Periodic Task will call the program at set intervals, every second as an example it's configurable.
- A program can also be called by an event. The event is activated using an instruction in ladder logic called the EVENT instruction. It can also be triggered by a consumed-tag, the point is it's configurable.

A task, therefore, is a program scheduler. A project can have only one continuous task and a varying number of periodic and event tasks. The number of tasks available is dependent on the family of processors you are working with. The project we are creating is a fairly small project containing a single program. The default continuous task is a sufficient scheduler. The thing to remember about the continuous task is that it has the least priority. In other words it is interrupted by everything, including HMI's and programming laptops.

Depending on what the overhead time slice is set at, the continuous task will be interrupted many times before completing, which is ok for this project, however on larger projects I would place the program in a Periodic task to ensure the code is executed before being interrupted by *everything* else.

Just be aware that a project containing multiple tasks will interrupt the continuous task to complete the other tasks. The least priority is always given to the programs in the continuous task, therefore you must program accordingly.

Program Sub-Routines

Routines belong to programs. Each program in a project will have at least one routine that is called when the program is run, by default this is called the *main* routine, however you can change the name to whatever makes sense. We could place all our Ladder Logic code in this single routine and name it *Everything*, however wouldn't it be nice to create several logical routines to make our code readable?

For instance what if we put the system run logic in a routine and called it *System Run* or *System Enable*? What about the emergency stop code, why not place it in a routine and call it *Emergency Stop*? As with the conveyors each could have a separate routine appropriately named.

I think you can start to appreciate how a project is developed. Routines are available to use at your discretion. At times they will be used strictly for code organization. At other times you can use them to selectively run code only when necessary to save on scan time. The fact is there are several reasons to break code up into routines. The examples here are basic and focus on organizing code into logical routines for both readability and portability and are very common.

Routines are made by right-clicking on the program and selecting add routine from the pop-up window. This is also where the type of routine can be selected. You can create a structured text, function block, sequential function chart or ladder logic routine. It is perfectly acceptable to create any type of routine; Ladder Logic can coexist with structured text or function block.

The Controller Organizer Window

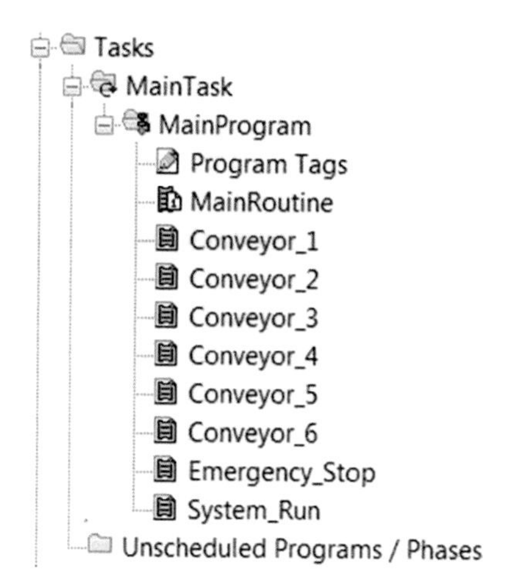

Take a look at the picture above, it tells a story. When routines are created they will show up in the controller organizer. The controller organizer is a hierarchal view of the project. Routine names are displayed below the program name just after the program tags. The main routine will always be the first routine displayed, after that routines are displayed in alphabetical order, not the order in which they are executed. Routine execution is configured using logic and instructions.

A quick glance at a well-organized project is all that is needed to determine how a project is laid out. The controller organizer is a map of your project. If you wanted to add an additional emergency stop station to the system you could find the routine that houses the emergency stop code and quickly add it to the project.

Code portability is also important when considering how routines will be created. Once the code for the first conveyor is complete it's a fairly simple process to replicate it for the subsequent conveyors.

One of the benefits of programming with Ladder Logic is that it can be the easiest language to troubleshoot when written well. The goal should be to make the code as easy to read as possible, within reason. If somebody who has never been in your program is able to quickly navigate and troubleshoot it you have done an excellent job putting it together.

Executing Code Using the JSR Instruction

Our code organization is complete. The right code will soon be placed in the right routines making the project readable and portable. The routines that were added will contain code, however the processor does not yet know how and when to execute said code. We know it will be done by routine, but how?

Every program contains one routine that is assigned as the program's main routine. The main routine is executed when the program is called. All other routines in the program must be called using a Jump to Subroutine instruction or JSR for short. For this project all subsequent routines will be called from the main routine, this is where all the JSR instructions will reside. Each JSR instruction will be placed unconditionally one per rung. The JSR instruction transfers the program scan to the called routine. Once the called routine is scanned the program scan is transferred back to the main routine directly after the JSR instruction. The first JSR instruction is executed, then the next and so on until all routines are executed. When the last instruction is executed everything starts over and the process begins again.

The JSR instruction can pass parameters and receive parameters. Although not used in this program, passing parameters gives you the ability to create less code by reusing the same code with different parameters.

For instance the program we are creating integrates a separate routine for each conveyor. The logic for each conveyor is identical, the only difference from one conveyor routine to another is the arguments associated with instructions. As an alternative we could have created this program using one generic conveyor routine. On the first call the parameters would include all the data to control Conveyor 1. On the second call the parameters would include all the data necessary to control Conveyor 2, and so on. Doing it this way would have required far less code and used considerably less memory. So why didn't I write it this way?

Sometimes it is ok to sacrifice one thing to gain another. While I could have made the code considerably smaller I opted to write the program using a separate routine for every conveyor. I did it this way to demonstrate tag scope, organization and portability. The biggest reason for writing the code the way I did was for readability, which I have emphasized throughout this book.

I have written code for systems similar to what has been demonstrated here. I have also had the opportunity to troubleshoot similar code written with the parameter passing method. Bottom line, the latter is harder to troubleshoot. The code is executed so fast it's impossible to see what is going on without creating watch windows and adding code traps. The parameter passing has its place, however for this specific example the individual routines are easier to read and troubleshoot.

Chapter 7 Questions

1. Every program will have one routine configured as the main routine. The main routine does what?

- The main routine interrupts all other routines in the program.
- All subsequent routines must be called from the main routine.
- The main routine cannot contain any instructions other than the JSR instruction.
- The main routine is always executed when the program is called.

2. True or false, breaking code into routines allows code to be organized, portable, and readable.

3. True or false, how routines appear in the controller organizer is how they are executed.

4. Which instruction is used to call subsequent routines?

- The AFI instruction.
- The JMP instruction.
- The JSR instruction.
- The SBR instruction.

5. True or false, when calling a routine a parameter must be passed.

6. How many Continuous tasks can reside in a PLC?

- It depends on the controller.
- As many as the memory allows.
- Each PLC can have one Continuous task.
- Each program can have one Continuous task.

7. True or false, a sub-routine in a program will call a task.

8. A Periodic task is best described:

- As a task that is run every time the Continuous task completes.
- As having at least one program that calls an Event task.
- As a task which is triggered by a configurable time schedule.
- As something that should only be used for alarming.

9. A Continuous task is best described as?

- The task with the least priority.
- Something that cannot be renamed.
- The default task and must be used.
- Containing all the programs in the project.

10. A program is a collection of____________.

- Routines, Tasks and Tags.
- Tags, Routines and Instructions.
- Instructions and IO definitions.
- All of the above.

Program 7 Answers

1. Every program will have one routine configured as the main routine. The main routine does what?

   The main routine is always executed when the program is called.

2. True or false, breaking code into routines allows code to be organized, portable, and readable.

   True, subroutines can be used to organize code for readability and portability.

3. True of false, how routines appear in the controller organizer is how they are executed.

   False, routines appear in the controller organizer alphabetically. With the exception of the main routine subsequent routines are called using the JSR instruction.

4. Which instruction is used to call subsequent routines?

   The JSR (Jump to Subroutine instruction).

5. True or false, when calling a routine a parameter must be passed.

   False, parameters can be passed to and from routines, however it is not a requirement. When parameters are passed a Subroutine (SBR) instruction is required.

6. How many Continuous tasks can reside in a PLC?

   Each PLC can have only one Continuous task.

7. True or false, a routine in a program calls a task.

False, programs are scheduled in tasks, however an EVENT instruction can call an EVENT task.

8. A Periodic task is best described as?

   A task which is triggered by a configurable time schedule.

9. A Continuous task is best described as?

   The task with the least priority.

10. A program is a collection of ***tags, routines and instructions.***

C-8 Adding Tags To a Project

Thus far we have created logic and organized it to make it both readable and portable. Now it's time to wrap up our logic by creating tags for the instructions. Thus far our conveyor logic has used five instructions: XIC, XIO, OTL, OTU, and TON instructions. All the instructions with the exception of the TON instruction require Boolean arguments that reference one bit of data, 1 or 0. Input instructions like the XIC and XIO instructions examine the value of the argument. Output instructions like the OTE, OTL and OTU instructions write a value to the argument. The TON argument is a group of arguments also called a structure.

Defining a Naming Convention

Before we define all of our tags we need to decide how they will be named. What naming convention will be used to make them consistent, organized and readable? Tags are organized in alphabetical order in the tag window editor just as routines are organized in the controller organizer. We need to consider how the tags will appear in the tag window editor. The key is to define the tag names consistently making them organized, portable, and legible.

The six conveyors will have abbreviated unique names C1, C2, C3, C4, C5 and C6 for conveyors one through six respectively. We could fully define the conveyor names, for instance CONVEYOR_ONE. Spaces are not allowed when naming a tag, the underscore is the accepted practice to use in place of the space character. Abbreviated names like C1 as opposed to CONVEYOR_ONE are more readable when used as arguments. Imagine a rung with multiple instructions. All the arguments have fully defined names. There is only so much space available to display the rung before the IDE will wrap the rung making it more difficult to read.

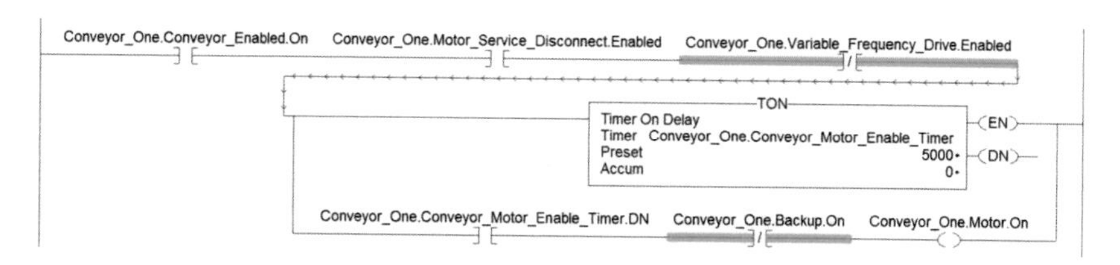

Now compare the same rung with abbreviated tag names below. Notice that the tags have members, they are User Defined Tags, which means each member also has a name. As each member of a tag is used for an argument the full tag name becomes larger, more illegible. Tag names should be as simple as possible while still being discernable.

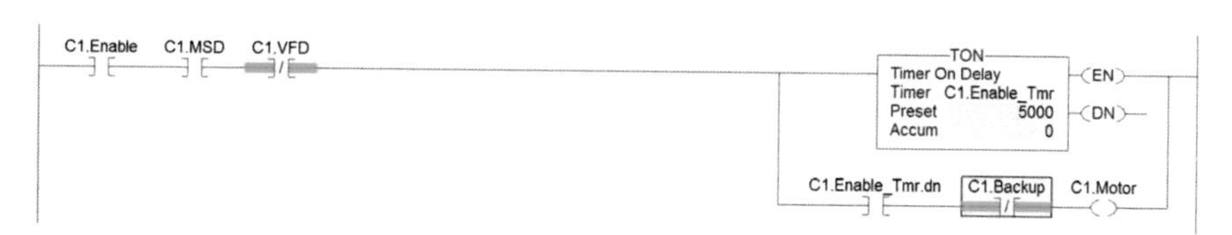

It make sense to use the same descriptor for creating all of our tags. The emergency stop buttons will read C1_EPB, C2_EPB, etc. The motor service disconnects will read C1_MSD, C2-MSD, etc.

Grouping the tags by giving them the same identifier creates consistency and allows all the different tags to be viewed together in the tag window viewer. What if we were to name the tags slightly different, that is put the descriptor at the end of the name? When viewed in the tag window the tags will be displayed by function, not by conveyor. All MSD inputs will be displayed together, all EPB inputs will be displayed together. Placing the unique descriptor in the front of the tag will insure that all tags pertaining to an individual conveyor will display together.

Name
⊞ C1
C1_EPB
C1_KSW
C1_MSD
C1_PE
C1_RPB
C1_VFD
C1_VOK
⊞ C2
C2_EPB
C2_MSD
C2_PE
C2_RPB
C2_VFD
C2_VOK

Neither organization method is wrong. For this project I want all the tags associated with a single conveyor to be displayed together. This way anyone can see all the variables that apply to a single conveyor. There are times where you may want all the descriptors at the end of the name. Be aware that tags should be defined with a consistent descriptor and the placement of the descriptor is important.

Defining Base Tags

Now that we have a naming convention lets define some tags. We will start out with the emergency stop buttons. There is one input button per conveyor. All buttons will have short abbreviated names, all capital letters, and a unique identifier. All emergency stop buttons will contain the abbreviation EPB which is short for Emergency stop Push Button.

Each tag requires a unique name and given that there will be six unique push buttons we must also give them a unique identifier. This identifier will be consistent throughout the tag definition process as described earlier. The names for the emergency stop buttons will be defined as follows.

- *C1_EPB*

- *C2_EPB*
- *C3_EPB*
- *C4_EPB*
- *C5_EPB*
- *C6_EPB*

This is done for all the input conditions for every conveyor.

- *C1_MSD*
- *C1_VFD*
- *C1_VOK*
- *C1_PE*

All the input conditions in the first rung of the emergency stop circuit are EPB push buttons. The tags need to be defined and associated with the XIC instructions. The tags will be defined as base tags, for now, with a data type of Boolean. The tag scope which I am confident you remember will be defined as program. The tag description will be kept generic and will read "Maintained E-Stop Button". While we are adding descriptions we may as well create a descriptive rung comment like "Emergency Stop Latch Circuit."

The emergency stop reset circuit has both the emergency stop buttons and restart buttons as input conditions. Following the naming conventions defined earlier the reset buttons will be named as follows:

- *C1_RPB*
- *C2_RPB*
- *C3_RPB*
- *C4_RPB*
- *C5_RPB*
- *C6_RPB*

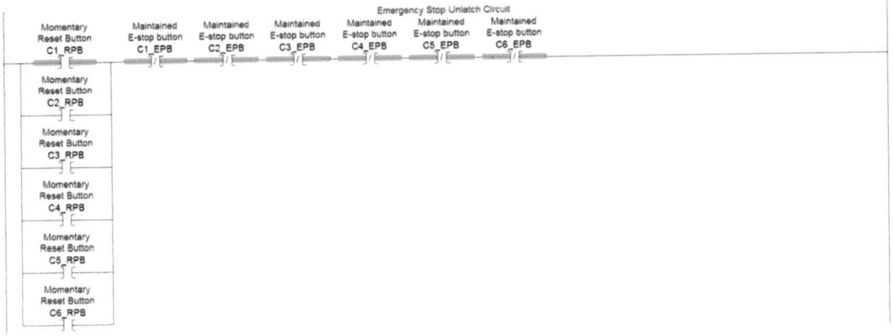

To finish up this circuit we need to create a Boolean tag for the OTL and OTU instructions. The OTL and OTU instructions will reference the same tag, one will set the data and the other will clear it. The name Estop should do the trick.

Organizing Tags by Using User Defined Types

Before we define any more tags, I think it's important to talk about how a microprocessor works. Why is this? Because defining a tag as a Boolean data type is not the best idea. I know what you are thinking, we just created six of them and now this is bad?

Yes, tags defined as Boolean are not optimized. Every processor in the world is driven by a clock which is nothing more than an oscillator. An oscillator will reliably oscillate between a high and low signal when a consistent voltage is applied. This oscillation is what drives microprocessors. A 32-bit processor will move 32 bits of data per every clock oscillation. A 16-bit processor will move 16 bits. As you know a Boolean tag consist of one bit of data, the value can be either 1 or 0. When a Boolean tag is updated the processor moves 32 bits with 31 of those bits null. A very inefficient and time consuming way to move data don't you think?

This is the reality of using Boolean tags. Now imagine you have thousands of Boolean tags in your program. I think it's fair to say that would be inefficient at best. If you had to move 1000 Boolean bits is would take 1000 clock cycles, but what if you could move those same Boolean bits utilizing all 32 bits per clock cycle? It would take only 32 cycles. Organization is the key here, you can write programs that take less memory, execute faster, are legible, and portable.

How do you organize your bits and bytes to accomplish all this? By using structures. All of the various languages have them; RSLogix5000 calls theirs User-Defined Types or UDT's for short. A UDT is treated like any other data type, with the difference being that you get to define the type.

Before we define the Estop tag consider the conveyor logic we created previously. When we laid out the logic we put descriptive names on the instructions, however we did not define the tags. I did this specifically because I didn't know how many tags or the type of tags that would be needed. I wanted to get some solid logic put together that would drive the process of creating the tags. Put another way, I wanted to develop logic to solve a problem rather than create logic from invented tags. For me the logic drives the development of the tags, not the other way around.

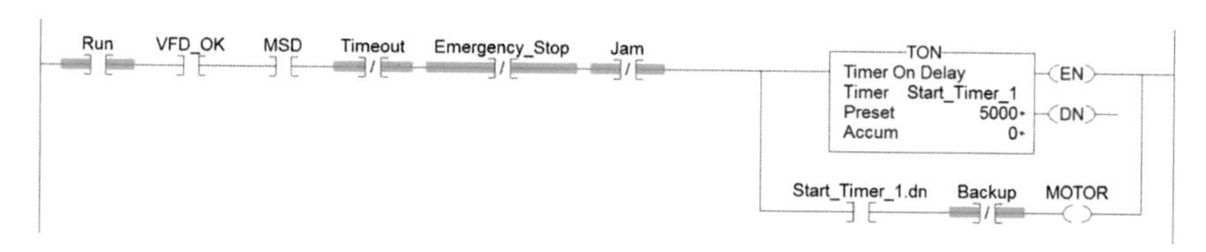

Take a look at the conveyor rung above. How should the various components be grouped? The *run* argument is not specific to an individual conveyor. The *run* argument enables and disables all the conveyors in the system. The estop argument is similar; when an emergency stop is activated all the conveyors in the system will stop. The conveyor start timer on the other hand controls a single conveyor, it is not used by the other conveyors in the system.

All the system tags will be grouped together in a structure and will be used by all the conveyors. The data type will be defined with the name *System* and will contain all the system members including the Boolean Estop tag discussed earlier.

The conveyor tags will be grouped into a structure defined with the name of *Conveyor*, a simple, descriptive name don't you think? There are a total of eight tags used on each conveyor rung and we have already grouped two of them into the System data type. We know our objective therefore the only thing left to do is come up with simple descriptive names and define the data type for each member of the conveyor group.

Optimizing User Defined Data Types

The creation of a member is the same as creating a normal tag outside of a UDT. All of the same data types are available for you to use. In the screenshot below you can see all the members of the conveyor data type. Now it's time to optimize this data type. Take a look at the members, there are three Boolean data types and two TIMER data types. The Boolean data types should be next to each other. Like data types are placed together in UDT's to optimize memory.

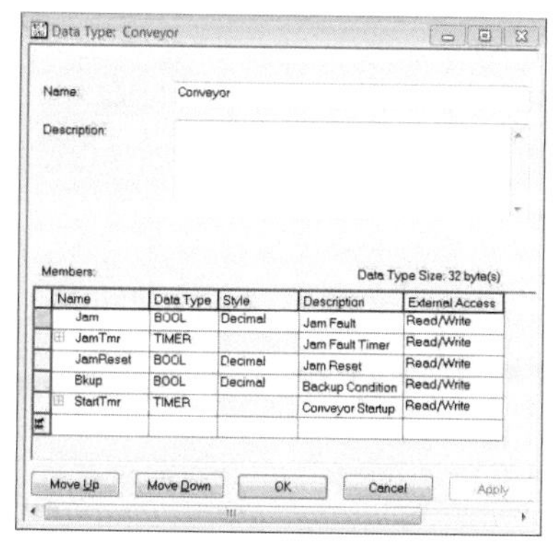

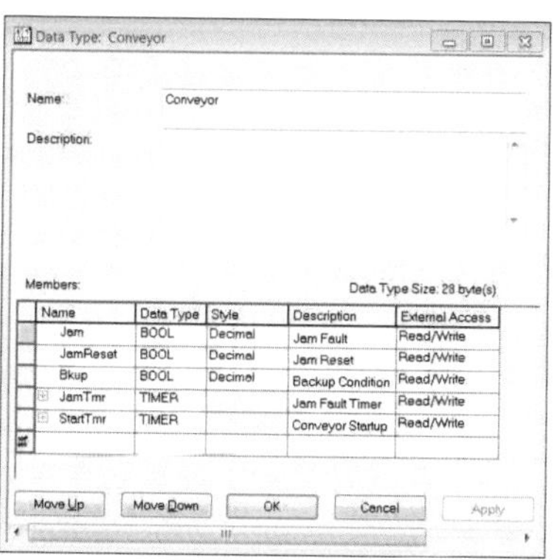

Look at the memory usage of the Conveyor data type in the first screenshot above. The data type consumes 32 bytes of data. There will be six instances of the conveyor data type, one for each conveyor. That is a total of 192 bytes of data. The second screenshot shows a reduction in memory usage by simply grouping the Boolean members together in the UDT. The total usage is now 28 bytes, which reduces the total bytes used to 168 bytes. Moving the members around is done by simply highlighting a member and using the “Move Up” and “Move Down” buttons. Doing this will not make much of a difference in performance with our little program, but imagine what it would do if you had 100 or 1000 conveyors. The point being that you should write all your programs the same way, by incorporating good practices you won’t have to worry about optimizing code somewhere down the line. Do this every time you create a UDT.

Following the naming convention we came up with earlier it should be easy to define all of our tags. There will be six tags defined as conveyor data types, and one tag defined as a System data type.

- **Tag Name   Data Type**
- C1   Conveyor
- C2   Conveyor
- C3   Conveyor
- C4   Conveyor
- C5   Conveyor
- C6   Conveyor
- Sys   System

More About Tag Scope

All of these new tags will be defined as program scope tags, but why? What is the real difference between program and controller scope tags?

Naming a tag and assigning it a data type isn't all there is to do when creating a new tag. The scope of the tag must be defined as well. Lucky for us there are only two possible selections to choose from. You could just wing it, after all you have a 50% chance of getting it right. You could define all the tags as controller scope, this would work. You could also define all the tags as program scope, this would also work. It seems either will work.

A quick review on program scope and controller scope tags is necessary. Program tags can be used anywhere in the program they are created. They cannot be used outside of the program in which they are defined. A project can have multiple programs. Controller scope tags can be used by all programs in the project.

Why not just use controller scope tags, that way tags from one program can be used in another, that's less code you have to write, correct? This is absolutely true. Some projects have multiple programs, and the various programs will need access to all the tags. If the tags were defined as program scope, logic would need to be written to transfer data from local tags to controller tags. Not an efficient way of doing it.

The cool thing about program tags is that it gives you the ability to create program scope tags that have the same name as a tag in a different program. So what? What does that do for you as the programmer?

Code Portability

Suppose we finish our program and we create all of our tags controller scope. The freight company we are building this system for gets an unexpected contract that will double the amount of freight processed in the warehouse. A new line identical to the one you just finished is already being installed. You need to get the program written yesterday, as is often the case. The quickest way to do this is to create a copy of the existing program and change all of the tag names. Each tag name must be unique. You can't have two controller tags with the same name, remember a tag is nothing more than a pointer to an address. You can't have two pointers pointing to the same address. Even if you could what good would it do, one address is enough.

If you used program scope tags you wouldn't have to change the tag names at all. The names would stay the same, remember program scope tags are invisible to other programs. In other words two different programs can have the same tag name. The tag C1 in program one is different from the tag C1 in program two. The tags point to different memory locations.

Using structures enables you to create a more complex group of tags once, you don't have to create tags and descriptions for every component in every conveyor. It was created once and used six times. This is the preferable way to define tags, this process uses less memory, is more organized, portable and easy to read.

Chapter 8 Questions

1. True or false, Tags are organized in the tag window editor alphabetically.

2. True or false. Using a Boolean data type is more efficient than using a DINT datatype when defining a tag.

3. Which words best describe the following statement? A User Defined Type can be used to ________ and _________ tags.

- organize, optimize
- minimize, utilize
- catalog, distribute
- describe, quantify

4. True or false. A UDT can contain a variety of different data types.

5. Describe at least one benefit of creating program scope tags.

6. Describe at least one benefit of creating controller scope tags.

Chapter 8 Answers

1. True or false, Tags are organized in the tag window editor alphabetically.

   True. If you want to keep tags together for easy access stick with a naming convention that will display all your various tags properly.

2. True or false. Using a Boolean datatype is more efficient than using a DINT datatype when defining a tag.

   False. The DINT is the most efficient data type, being that it is a 32 bits and the processor is a 32 bit processor. The DINT data type is natively optimized.

3. Which words best describe the following statement? A User Defined Type can be used to ***organize*** and ***optimize*** tags.

   Grouping Boolean tags in a UDT will optimize them as opposed to creating them individually outside of a UDT.

4. True or false. A UDT can contain a variety of different data types.

   True, a UDT can contain all the native data types as well as nested UDT's.

5. Describe at least one benefit of creating program scope tags.

   Other programs can have the same tag name, which allows programs to be portable.

6. Describe at least one benefit of creating controller scope tags.

Every program in a project can access the tags as opposed to a program scope tag.

C-9 Input & Output Modules

As you may have guessed a PLC is all about control. A PLC accomplishes this by reading inputs, executing logic and switching outputs. The *logic* must connect to the outside world of buttons, switches, motors, sensors, robots, cameras, and the like. Input modules connect real-world-inputs to logic. Output modules connect logic to real-world-devices. As an example take a look at a basic modular setup comprised of a rack, power supply, PLC, input module, output module and communication module.

Individual modules can be placed anywhere in a rack. The module with the key switch is the PLC, to the right of the PLC lies a 32 point DC input module, 32 discrete input signals are available to connect to real-world-devices. To the right of the input module is a 32 point output module where devices like motors, valves and relays can be connected to interface with the program.

The PLC and various modules could be placed in any position in the rack, it really doesn't matter where any of them are placed. On the very end is a 1756-ENBT module which is used for Ethernet communication. This is where Ethernet devices can be connected. I use this port to connect my programming laptop to the PLC to write the logic that will interface with the input and output cards.

This module can also communicate with HMI's and other various equipment.

All of the various PLC control hardware lies in a rack. The whole gang is there: the PLC, communication card, input card and output card. In order for the PLC to know what devices are available and where they reside in the rack an IO configuration map must be set up.

IO Configuration Tree

This is fairly straightforward process, the controller organizer has an area called the IO Configuration. This is where the components of the system are added to the project with the exception of the controller and the rack. The controller and rack are selected when a new project is created. Check out this link for a quick video (www.ladder-logic.com/hello-world/) on how a processor is selected from the IDE. I have also put together a free manual that covers adding modules that is available at the beginning and end of this book.

Let's start with the 32-point input module. The module is defined as an Allen Bradley IB 32 input card. To add components to a rack, right click on the rack in the IO tree. A window will appear and a *new module* selection will be available. Once a new module is successfully added and configured, module-defined tags which reference the new IO are available at the controller level (controller scope) for use in logic. The tags are accessible by all the programs in the project. These tags are the addresses to the 32 discrete inputs. This is where the real-world-inputs such as a sensors, buttons and switches interface with programmable logic. The input devices set and clear the variables. In the case of an output module the associated tags are set and cleared using instructions like the OTE and OTL instructions.

When a module is added it can be configured. For example the IB/32 input module has configuration options for change of state

(COS) and filter times. This data is stored in module-defined controller tags. For our purpose we are concerned only with the input tags. The same goes for the output card in respect to the output tags.

The software will name all the IO tags for the modules that reside in the rack that contains the PLC. The tags will contain the name *Local*. A project can have multiple racks with multiple modules, however our example contains one rack with only a handful of modules. All other tags associated with modules that do not reside in the Local rack will take the name of the module. The module name is created, by you, when a module is defined.

Module Defined Tags

Let's break down the tag names to understand what they mean. The 32-point input module has 32 discrete inputs that can be accessed by the Ladder Logic. The module-defined tags we are concerned with are **Local:1:I.Data.0** through **Local:1:I.Data.31**.

Notice that the numbering starts with .0 and ends in .31. It is common in programming to start with zero as the first addressable subscript. Counting zero as the first subscript means that thirty-one is the last addressable subscript. Zero through thirty-one gives us thirty-two points, don't believe me count them yourself. The tag **Local:1:I.Data.0** can be used as an argument for the XIO instruction.

Module Defined Tag Breakdown

- **“Local”** indicates the module resides in the rack containing the controlling processor.
- **:1:** indicates the slot number where the module is located. The slot numbers begin with zero. Zero is the first slot.
- **:I.** indicates that this is an input, if it were an output it would read **:O.**
- **.Data.0** indicates the discrete point.

Chapter 9 Questions

1. True or false. A PLC must reside in logical slot '0' of a rack.

2. IO modules are added to the project by right-clicking on a rack in the ____________.

- Controller Tag Editor
- Controller Organizer
- IO Configuration area
- Program where they will be used

3. True or false. When a new IO module is added you must then create tags that address the module.

4. The address MCP_3:4:I.Data.12 points to an input that resides where?

- Four slots to the right of the PLC
- Four racks from the rack that contains the PLC
- In slot 4 of a remote rack
- Nowhere, the address is invalid

5. An analog output card residing in slot three and the same rack as the PLC will have tags that look most like ________.

- Analog:3:O.Data
- Output:3:A.Data
- Local:3:O.Ch2Data
- None of the above

Chapter 9 Answers

1. True or false. A PLC must reside in logical slot '0' of a rack.

   False, the PLC can sit in any open slot. As a matter of fact several PLC's can reside in a single rack.

2. IO modules are added to the project by right-clicking on a rack in the ***IO Configuration area.***

3. True or false. When a new IO module is added you must then create tags that address the module.

   False, module defined tags are available for use after an IO module is added to the IO Configuration. It is then possible to use the tags as arguments in your ladder logic.

4. The address MCP_3:4:I.Data.12 points to an input that resides where?

   In slot 4 of a remote rack. If the address was named "Local" the module would reside in the same rack as the PLC in slot 4.

5. An analog output card residing in slot three and the same rack as the PLC will have tags that look most like ?

   The correct answer is Local:3:O.Ch2Data. Analog inputs and outputs have channel data.

C-10 Tag Types

We have covered data types quite extensively, for instance a tag defined as a DINT will address 32 bits of data in the processor memory. What we have not covered is the tag type which is different than the data type. There are four selections to choose from when defining a tag type: Base, Alias, Produced and Consumed. So what do these mean?

The Base tag is a generic tag, there is nothing special about it. Most tags you create, use and troubleshoot will be base tags; they behave just as you would expect. All the tags we have created thus far have been base tags.

Base, Alias & Produced/Consumed Tags

- A **Base Tag** is a generic tag, nothing special about it, it's the most commonly defined tag.
- An **Alias Tag** is a tag that references another tag. It's a way to use two different tag names that point to the same address. I use Alias tags to make my code more legible.
- **Produced and Consumed Tags** are tags that communicate with peripheral devices, such as another PLC. Produced & Consumed tags can be used to transfer data between two devices. As an example produced data from produced tags in one PLC can be read with Consumed tags in another PLC.

Using Alias Tags to Point to Module Defined Tags

The tag ***Local:1:I.Data.0*** is an address to a discrete data point. Interpreting the tag data tells us that the data point is an *Input* and belongs to the module in *slot 1* and the module is housed in the same rack as the controller. Keeping with the standard of making our code as legible as possible it would be nice if we had a more descriptive tag name then an address to a data point.

What if the input tag ***Local:1:I.Data.0*** is an address to a reset push button? This is where tag aliasing can be effectively used to create legible code. The module-defined name cannot be changed to something more descriptive, however the tag can be aliased to a descriptive tag name like C1_RPB. Whenever the momentary push button wired to the input ***Local:1:I.Data.0*** is pressed the value of the tag C1_RPB will be true. When the button is released the value will be false. Notice how the tag looks when it has an alias. The tag name is C1_RPB and the alias is **<Local:1:I.Data.0>**. An alias address is surrounded by the < > characters.

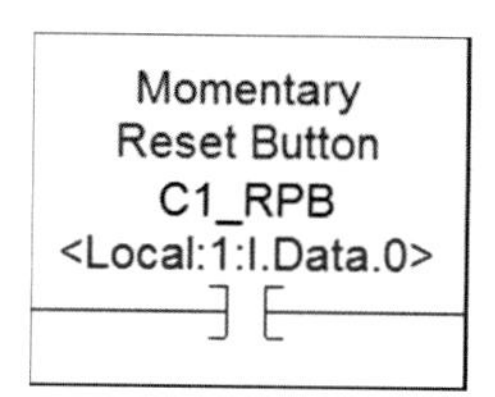

All of our input and output tags created earlier were defined as base tags, however now that the IO has been configured we can change the tag from a base tag to an alias tag and associate the proper address to the tag. Obviously schematics will need to be developed to show where all the various inputs and outputs are landed (*hint, creating electrical schematics is part of the job of a controls engineer, it's not all programming*).

Produced and Consumed Tags

Controllogix processors support what is referred to as the producer consumer model. Put simply this is a way for two or more processors to share information. A produced tag sends information over a network or the backplane for another processor to read, utilizing an appropriately configured consumed tag.

There are a couple of rules to note when using produced/consumed tags. Produced and consumed tags must be of the data type DINT and a consumed tag must be associated with a producer (another PLC).

I'm not going to extensively cover the producer consumer model here, but suffice it to say that it is available. I mention this because I know you will see the option when you create tags, and naturally you will want to know what it means. If you come across a produced or consumed tag while troubleshooting it's good to know that the data is being broadcast to a consumer or being consumed from a producer usually via a network connection. The communication is set up similar to how IO modules are added. The IO tree will always contain the producer in it.

Chapter 10 Questions

1. I/O modules can be of what type?

- DC Input modules.
- AC Isolated relay output modules
- Analog input modules
- All of the above

2. True or false, the PLC must be placed in the first logical slot.

3. An alias tag refers to a tag that.

- Represents indirect addressing.
- May or may not be wanted.
- Is called by the JSR instruction
- References a tag that has already been defined.

4. A Consumed tag refers to what?

- A tag that has already been used.
- A tag that receives data from a producer.
- Cannot be used in ladder logic.
- None of the above.

5. Modules are added via the ____________.

- IO assembler window
- IO configuration tree
- Configuration software
- Definition files

6. The producer consumer model is best described how?

- Programs consume produced routines.

- How routines are compiled as producers and programs consumers.
- A tag based communication method between processors.
- Describes the IO module relationship.

Chapter 10 Answers

1. I/O modules can be of what type?

   All of the above. I/O refers to Input Output.

2. True or false, the PLC must be placed in the first logical slot.

   False the processor can be placed in any slot in a rack.

3. An alias tag refers to a tag that.

   References a tag that has already been defined.

4. A Consumed tag refers to what?

   A tag that receives data from a producer.

5. Modules are added via the ***IO configuration tree.***

6. The producer consumer model is best described how?

A tag based communication method between processors.

C-11 Finishing the Code

Now that all the tags are fully defined it's time to add the defined tag names and copy the code from the first conveyor to the remaining five. This can be done a couple of different ways.

For the first conveyor routine I will simply add all the necessary tags to the instructions one at a time until complete, which is a simple enough process. I do the same thing for the conveyor run and emergency stop routines.

The same thing could be done with the remaining five routines, however that would be time-consuming and leave us wide open for mistakes.

Replicating Code Using the Manual Copy and Paste Method

One way to copy all the code from the first conveyor routine to the next would be simply to copy all the rungs from the completed conveyor and paste the code into the second conveyor routine. Once the code is pasted it would then be necessary to change all the tag names from C1 to C2.

This process works fine when there is a relatively small amount of code to replicate. The process can be sped up a bit using the search and replace capabilities of the software. Similar to common text editors, the ladder editor employs a configurable search and replace utility. Be sure to select current routine only or the software will replace the tag names in the first routine as well, I have done this more than once.

The hot-keys CTRL C and CTRL V allow you to quickly copy and paste code. Simply highlight the instruction, rung or group of code you want to copy then use the hot-keys or right click and select copy from the pop-up selection window. Do the same with the paste function, simply highlight the area you want the copied code to land and paste it.

The Copy and Paste Method By Routine

The second method essentially does the same thing only a lot quicker. You can simply highlight the first conveyor routine and either copy the entire routine and paste it or drag the routine up to the program in the controller organizer and drop it. Either way will duplicate the entire routine with an appended name. The name would then be changed from Conveyor_1 to Conveyor_2 and all the tags inside the subroutine would need to be changed from C1 to C2, again using the search and replace utility.

This is a quick and effective way to duplicate a routine without replicating the code instruction by instruction. I use this method when I want to quickly duplicate a routine without creating new tags.

Using Search and Replace While Online

There will be times when you will be required to add duplicate code while online. The copy & paste method works fine for this, however the search & replace utility does not function when online. In order to get around this it's possible to copy the code and paste it into a simple text editor like notepad or word. Use the search & replace utility in the text editor to replace the tag names, then copy the code and paste it back into the text editor. To do this create a blank rung in the editor and double-click it or press the enter button. At the top of the ladder editor a rung text editor will appear. There is a drop down window that will allow you to change the input from ASCII to Neutral text, select Neutral text and paste the code into the window to the right.

This little trick has saved me a lot of time when duplicating code that must be done online. Changing tag names multiple times in multiple areas is tedious and therefore prone to errors. Time is big, but consistency is key here.

The Import/Export Method

Using the Import/Export utility enables an entire program to be exported then imported back into the project. The beauty of doing it this way is that tags can be created and the IO alias can be changed making the entire process as accurate, quick and automated as possible. I can literally add another conveyor system to the project complete with the correct IO in minutes. None of this works properly however, if you don't create programs that are modular.

It is also possible to import and export Routines, Add-On Instructions and data types.

Exporting a Program or Routine

The import and export utility is easy to use. Simply highlight a routine, program, Add-On instruction or data type and right click it. From the pop-up window select *export routine, program etc.* A window will appear that enables you to name the file and select the directory where the file will be stored. The file will be saved as an .L5X file which is nothing more than an XML file. It is possible to open and edit this file with an appropriate web browser or text editor.

Exporting a file is simple, I suggest placing the exported files in a project folder, the same one that houses the project you are working on to keep things nice and organized. The real magic happens when you import the file back into the project.

Importing a Program or Routine

As a word of caution do not import a routine or program to an online processor, in my experience it never works out. The application will crash. Importing a routine or entire program is best done while not online with a processor.

To import a routine right click on the program in the controller organizer that you want to import the routine into. From the pop-window select *Import Routine*. A window will appear that allows you to select the directory where the L5x file is located. You can also change the type of file to import and where the file will be imported to.

The import configuration window will appear. This utility has a host of operations that can be selected. When importing the conveyor routine to create a new one I will change the final name to the new conveyor name. I then select Tags from the Import Content window and change all the C2 tags to C3. I then change the C1 tags to C2 and import the routine. My project now has a new routine appropriately named with all the correct tags in the correct places.

On larger projects with more tags I will use the search & replace utility at the top of the import configuration utility. The utility has several options like create new tags or use existing. This is all configurable. For instance, if you have not previously created tags this utility will create new tags for you. This little utility helps me replicate error-free code quickly. It takes some practice, but once you master it nobody will be able to replicate accurate code faster than you, that I promise. Just remember to save your project before you import especially if you are not sure what you are doing, that way if the results you were looking for are wrong, it's not a big deal, just open up the last good copy. These things take practice don't shy away from using them they will make your life much easier. Just remember to save a copy beforehand.

Create Modular Code

Suppose for a second that you were asked to create an identical conveyor system controlled by the same PLC. How would you do it? You could create new routines and copy all of the code. Of course you would have to create new tag names and routine names for all the new code. It would be possible, but it sounds like a lot of unnecessary work to me.

What if you duplicated the entire program? You could take an exact copy of the first program and create a new one with a different name. Because you created all of your tags as program scope you wouldn't have to rename a single tag. The same would be true for the routine names. You made your code modular so you can now plug it in anywhere and use it. Ninety percent of your job is complete all that is left to do is add new IO modules to the project and alias the new IO tags to the correct addresses.

Modular programming just saved you from copying, pasting and rewriting code. You already went through all the steps to commission a system, you found and corrected all of your mistakes, why do it all again?

You would not believe how many programmers do things the hard way. Tags are created without much thought about scope; some are local while others are controller. Rather than breaking the code into routines it's all in one or two routines. Sometimes it's in several programs when it should have been in one. It's easy to spot a program written without much thought given to the basics like tasks, programs, routines, and tags.

A single project can have multiple programs in multiple tasks. It's up to you to decide how the code is laid out and executed. Do not use multiple tasks because you can, use them because it makes sense. The same goes for programs, routines and tags. If it makes sense to make all the tags controller scope, do it.

A Thought About Programs

Now that you can see the power of modular thinking I would like to cover exactly what a program is and is not. A program is a capsule that contains tags, routines, and instructions. A program is not a project, but rather a component in a project and this component should be modular.

Create Your Own Code library

There is a good chance that somewhere down the road you could use the conveyor logic from this project for an entirely different project. If your code was in a single routine or grouped poorly you would have a difficult time finding and extracting what you need. It might actually be easier to start over. If you organize your code properly you have the beginnings of your very own code library.

Create a code library; all programming languages make use of libraries. Libraries are nothing more than code that has already been developed that you can take advantage of. There really is no reason to reinvent the wheel every time you create a new project. Make libraries of your code, this is possible because you made your code portable and you documented it well. A library can be whatever you want it to be. It can be several projects, several routines or programs. The point is to make copies of all the projects, programs and routines you have worked on and place them in well-organized folders with supporting documentation. Create a word document that explains why you decided to hold on to the code. Take for instance the conveyor code we have been working on. What if you work in a water treatment facility for the next three years, but then find yourself working on a freight handling system. Rather than starting from scratch you can browse your library, read your documentation and start your new project with a bang.

C-12 Wrapping It Up

All of the logic is written the tags have been defined and associated with instructions. Tag names were purposely kept short yet descriptive to keep the ladder editor from wrapping rungs around on themselves. The logic was strategically placed in routines to make the code legible. The routines were appropriately named so that any programmer is able to read the project. IO was added to the project and the module-defined tags were aliased to tags that were descriptive. It sounds like this project is ready for the books; time to wrap it up and call it a day, right? Not quite yet, we still need to talk about one very important thing.

Document Your Code

Thus far I have not written much about code documentation. Adding comments to code should not even have to be mentioned, however I know there are some who feel that comments are unnecessary, and you would be in the minority my friend. A simple rung explanation would mean the world to the next guy. Be kind enough to comment, tell us what you were thinking. Now you don't have to go crazy with the comments, a simple descriptive comment telling the world why the logic is there will suffice.

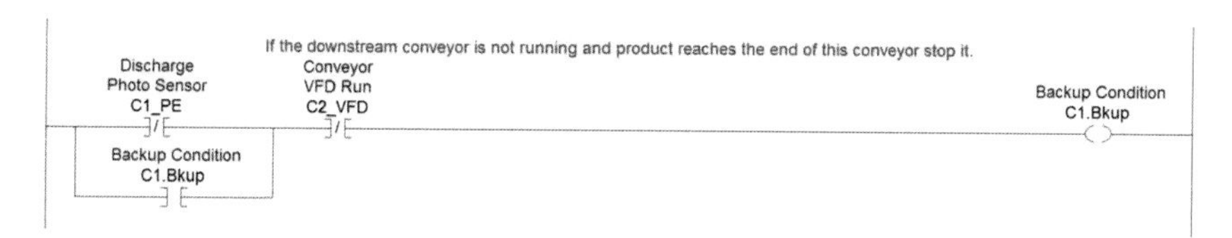

How and Where to Add Comments

Comments can be added to tags when they are created or they can be added later by right clicking a tag from the ladder editor and selecting *properties*. Tag descriptions can also be added from the tag editor as well.

Rung comments are added by right clicking a rung and selecting *Edit Rung Comment*. A rung edit window will appear where you can simply type in the comment. Alternatively you can use the hot key combination CTL-D when the rung or tag is highlighted from the ladder editor.

What Makes a Good Comment?

A common mistake is creating comments that detail how the logic works. It usually goes something like this "The XIC argument when high, will set the OTE argument". No kidding, I can figure that out myself. Instead explain why you put the rung there in the first place. Something useful like, "This is the door interlock circuit" or "This is the cup height measurement circuit". You do not need to give a play by play of the logic, the logic is easy enough to figure out, tell us what purpose the logic serves.

It is not necessary to comment on every rung however complex rungs deserve a good explanation. For instance if there are four rungs that control a height inspection process and a comment on the first rung gets the point across no further documentation is necessary. Perhaps the comment reads: "This is the cup height logic, the cup must not be above or below a set limit". This explains the intention of the logic, what its purpose is. Explaining how every rung works adds no real value. Rung comments need not explain how the code works, they should explain why the code it there.

Tags can have descriptions as well. In many cases the tag's name may be description enough, however in some instances a descriptive comment may be necessary. As an example the Flix Elixir machine has several operating modes. The tag FlixElixr.Mode is used in the code as a parameter in a comparison instruction.

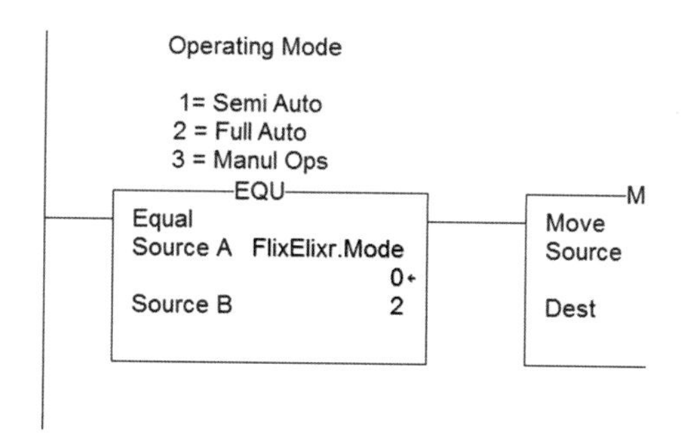

The tag description explains it all. Without the description you would have to do a lot of searching to figure out the meaning of the different tag values.

Using descriptions is not an excuse to write illegible code. The more legible your code the less descriptive you can be with your comments. Add rung and instruction comments to enhance the readability of your code don't rely solely on comments to tell the story of the logic.

Wrap Up the Project

The information in this book was written for people who are looking for a better understanding of how to write and troubleshoot Ladder Logic. In order to understand how the logic worked you needed to understand some basic components of the language like variables and instructions and how they relate to one another.

You the reader should now be able to open any Ladder Logic project and quickly discern how many tasks and programs are in it. You should also be able to drill down through the controller organizer and find out how many routines are in a program and what each routines purpose is, that is if the program you are looking at was written legibly. You should also be able to identify global tags and program tags as well as know the difference. Knowing the difference between the two allows you to effectively read the logic.

When it comes to reading Boolean logic you should have it nailed. You know how to differentiate between input instructions and output instructions. You can identify AND, OR, NAND & NOR logic and create similar logic which you can use to make the next Flix Elixer machine, whatever that is. When you come across an OTL instruction you now know that the tag associated with the instruction will remain high until set low with an OTU instruction. You also know better than to use the same tag with more than one OTE instruction. Put simply, you have the basics down.

You can also identify all the various modules that make up the project with a quick glance of the IO configuration tree in the controller organizer. You can identify modules and where the modules reside. You can find the discrete IO points of these modules and trace them back to where they are used in the code.

What if you come across a Consumed or Produced tag? You know that the Produced tag is sending data to another controller located somewhere in the IO tree, and a Consumed tag is receiving data from another processor. What about the Alias tag? You now understand why the tag looks so different from the normal base tag, and furthermore you understand why. Where would you use an Alias tag?

An excellent next step would be to open an existing project and study it. You will have a fresh insight as to why the project was organized the way it was. I encourage you to open a random project and look it over, dissect it rung by rung. Ask yourself why the programmer or programmers created the project the way they did. Doing this will give you more insight then a hundred books.

The intention of this book was not to give a step-by-step instructional on how to do everything. In my experience a book like that is overwhelming and boring at the same time. In the end books like that are less than helpful. You cannot successfully master programming by reading books alone, you actually have to get your hands dirty. This book is meant to be your compass, a road map to get you started in the correct direction. It gives you the big picture, it is up to you to grab a copy of the software and experiment with it, you need practice. Create some tasks, programs, routines and tags, get familiar with the IDE, and learn its idiosyncrasies. You have all the tools necessary to create logic, go do it.

You know all the basics necessary to become a successful programmer so what are waiting for? There is an undeniable demand for qualified technicians and engineers, the jobs and the money are out there go get it.

Your Free Gift.

I created a manual that details the Graphical User Interface of the Logix5000 programming and configuration software. This free manual compliments this book. The manual covers key areas such as the controller organizer, tag editor and IO Configuration tree. It's free and exclusive to those who have purchased my book.
www.ladder-logic.com/book-free-gift/

www.ladder-logic.com/book-free-gift/

Bonus Chapter: Common Procedures and Troubleshooting.

This book would not be complete without a chapter about some of the common troubleshooting techniques and invaluable tools I use every day.

Whether you are in school, maintenance, or the engineering field the following information will be useful. At some point you will find yourself needing to force IO, download or upload a project, make online edits and cross reference variables. I created this chapter to help you, think of it as an introduction to some of the common tools and procedures you will be using, in many cases daily.

Input and Output Forcing

Forcing allows you to override IO values. If the logical result of code turns an output off it is possible to force the output to an on state. It is an excellent tool for troubleshooting. For example, a pilot-light controlled by a PLC is not turning on when it should be. Examining the logic reveals that there are several conditions that must be satisfied in order for the light to turn on, however maybe the light bulb is burned out, or a fuse is blown. By forcing on the output you can quickly determine whether the light bulb and associated circuitry are functioning properly.

Scenario 1

Simply force the output on. Did the light turn on? If so then the light bulb and associated circuitry is good and there is some conditional input that is keeping the light from turning on. Look at the rung and use the process of elimination to find out why the light is not coming on.

Scenario 2

The light bulb did not come on after the output was forced. Check the light bulb, is it burned out? If the light bulb is good there is a problem with the circuit. It could be the IO module, the light socket, the power supply, fuse etc. The point is that forcing the output on effectively and quickly diagnosed the problem.

Inputs and outputs can be forced either on or off. Forcing IO is a great troubleshooting tool and can be used to keep a piece of machinery running while waiting for parts. Always use caution when forcing IO. Never bypass safety circuits.

Safety circuits are designed in such a manner as to not allow a simple force or similar action to bypass the system. Use common sense and remember that a force is meant to be temporary. A simple rule is to use a force for troubleshooting not a permanent fix.

IO Forcing Examples

The following are some real world situations where I have used forces for troubleshooting purposes.

By forcing an output high I discovered that the output on a small Micrologix PLC had gone bad. The circuitry inside the PLC that controlled the output was no longer functioning for whatever reason. The solution for the customer was to move the IO point to a free output rather than replacing the PLC. If you do this make sure to use the cross-reference tool to find all the instances of the IO point.

Commissioning new projects can be a challenge. More times than not things do not go as smoothly as they do in the lab. Every single time I have commissioned a system I have to wait for components to be wired. In order to test components of the code I have forced inputs on or off while waiting for the rest of the system to be wired. The code is looking for interlocks that are missing because peripheral equipment is not yet online. Forcing the interlocks on/off allows me to commission and test parts of a system.

IO forces must be enabled in order to activate them. They are enabled from the online toolbar or the menu bar by selecting *Logic -> I/O Forcing*.

Using the Cross-Reference Tool

Back in the day if you wanted to run a cross-reference on tags you had to go through an actual paper printout made from a dot matrix printer. The cross-reference print out could fill an entire box. Post-it notes came in handy, as I used them to mark the various places where variables were used in the logic, it was time consuming and slow.

Thankfully technology has changed PLC's for the better. Now cross-referencing is done with the click of a button. All instances are displayed and hyper linked. To use the cross-reference tool simply right click a tag, routine, or module. Select Go to Cross Reference from the pop-up menu. A cross reference window will appear displaying all instances of the tag, routine, or module. Alternatively you can use the hot key combination "CTL-E".

The cross reference window will display all instances of the element, the exact location used such as program, routine, and rung number. All instances are hyperlinked; double clicking one will navigate the software to where it is used in the ladder editor. Once you navigate to the ladder editor you can return to the cross reference window using the hot keys "CTL-TAB "or from the Windows Menu (Top of the Screen) select Cross Reference.

The cross reference tool is indispensable. It can be used to quickly navigate anywhere an element is used in the application. You will find that this one tool is used more than any other. As an example pretend you have written code for a new machine. You get a call from maintenance claiming your code has reprogrammed itself. I'm not kidding this happens, well kind of. After some troubleshooting you trace the problem to a faulty input local to an IO Module. You need to get the Clod-Hopper-Infuser machine back up and running quickly, so you make an executive decision and land the input wire on a different input. Just one problem, the project is big, really big, the input is used in multiple locations. Luckily you know how to use the cross-reference tool, so you do it, and find the input is used 37 times in 6 different programs. A simple cross reference, then address replacement and just like that the Clod-Hopper-Infuser is back up and doing whatever it does.

The Controller Key Switch

All Logix5000 processors have a key switch located on the front of the controller. The key switch is a three position maintained switch and allows you to put the processor in one of three modes. The first mode is RUN, leaving the key switch in this position will allow the processor to execute tasks programs and routines. It removes any remote control of the processor. If for instance you wanted to download a new project or do some online edits this would not be allowed while the key is in the run position.

The next position is the REM or remote position. Placing the key switch in this position allows the processor to receive online edits, downloads and the like. When you download a project in remote the software will ask to change the processor mode to program, and after the download you will have the option to put it back in run from the software. The remote position allows you the ability to interact with a project that is running in the processor. The tasks and programs will behave the same as when in the run position, however you will have access to manipulate the process.

The last position is PROGRAM. When placed in this position all tasks are disabled. The project in the processor is not executed. In this position the processor will accept a download as well as online edits, without the project being scanned.

Uploading and Downloading Projects

Knowing when to upload a program and when to download a program is extremely important. Downloading a program consists of transferring a project from a terminal such as a laptop into a PLC. Uploading a project transfers it from a PLC to a programming terminal such as a laptop. This is a simple yet extremely important concept.

I have been witness to novice programmers mixing this up. What they really wanted to do was get online with a PLC that was running production, however the project residing in the laptop in this case was not exactly the same as the project running in the PLC. When the technician connected to the PLC to go online the software flagged the discrepancy and prompted the technician to either upload or download the program. The technician then decided he would download the program thinking that the program would download from the PLC to the laptop when in fact the program went the opposite way from the laptop into the PLC.

The real problem in this case was that the technician had an older version of the program which was downloaded into the PLC and as you can guess the equipment did not operate properly after the download. In this case the technician was completely out of luck because he had just overwritten the working program with one that did not work. He had no Backup of the correct program to download back into the PLC. As you can imagine there were a lot of unhappy managers calling lots of meetings.

This is a real case scenario, it happens all the time. In this story the correct code was eventually found and downloaded, however it's not always the case. There are literally tens of thousands of PLC's running programs in industries throughout the industrialized world. Many of these PLC's were programed years earlier by people who have moved on, contractors, maintenance technicians, engineers, etc. Many of these places have yet to employ full time controls personnel. Backup copies of programs may very well not exist.

Special Considerations

There are other things to consider before downloading. Controlnet for instance can be affected by downloading a project. Controlnet is a network used for communication and control of peripheral I/O. Controlnet configuration data is stored in a PLC. If the Controlnet configuration is different in the project that gets downloaded from the actual configuration all communications on that Controlnet will be disrupted. Some of these larger systems can have several Controlnets attached to thousands of I/O points. Make sure to proceed with extra caution when downloading to systems that utilize special networks.

As a word of caution before you download to a controller first upload a copy of that program and save it, save it, save it. Make this something you do every time, no exceptions. Then, and only then, download your new project. If everything goes to hell after that simply download the project you know to work. It's a life saver, really it is. Seriously make this a practice you do not skip, ever. I don't care how confident you are in your programming abilities, it takes very little time to upload and save a project from a PLC that is currently running a production line, sort line, feed line, or whatever. What you're doing is essentially making yourself a save point similar to a video game. Who wants to go back and do an entire level again? No one, I can't overemphasize this one too much, always upload a copy and save it with a unique name before you download over a program that is currently running a process. You can thank me later.

Uploading a Project

You can connect to a PLC and upload a project from the PLC's memory to a programming computer if that program has not been locked out. It's possible and in many cases security measures are taken to keep people from uploading or downloading projects. In your case however you will be the engineer/technician who has access to the PLC to upload and download.

If you have a copy of a project on your computer and connect to the PLC with the same project you can simply go online with the PLC. However if the copy of the code on your computer is different from the copy in the PLC you will need to upload the project from the PLC. It's possible to upload the project from the PLC to the computer without a similar copy. You simply upload the project and give it a new name or keep the default name of the project in the PLC. Once the program is uploaded it is possible to go online with the PLC for troubleshooting, code changes, or just monitoring. If you upload a fresh project without having a copy of the program you will not have any descriptions in your code. Descriptions do not get downloaded into the PLC, they stay on the programming computer. All the hard work someone went through to put descriptions in the project will be lost.

Pro Tip: If you have a copy of a project, even if it is an old copy, upload the new project to the old project on your computer. Doing so will merge the file you upload from the PLC into the file local to your computer. This way you can have the descriptions that were in your local copy merged with the program running in the PLC making your job just that much easier.

If you have a project that has no descriptions then you probably have a case where someone, maybe yourself, uploaded a program from a PLC without merging the uploaded file with a copy that has descriptions. If that is the case find an old copy of the program and upload the PLC project to the old file copy to get your descriptions back. I have yet to see a program written that has no descriptions.

Most, if not all, programmers will add some descriptions to the code, some more than others. If you cannot find a copy of a program containing descriptions then you may have to add your own, but it will be worth it, and it's a great way to learn the various components that make up the program.

Downloading a Project

Downloading a project is fairly straight forward. The first thing to check before downloading a program from a computer to a PLC is that the program does not contain errors. Programs with errors will not download to a PLC for obvious reasons. To see if a program has errors from the menu bar select Logic -> Verify -> Controller. This will verify all the code in the project. If there were no errors the program is ready to be downloaded into a PLC. If the program has errors an error window will display the errors. The errors are listed in the error window and are hyperlinked to the actual error. Which means clicking on the error will navigate the ladder editor to the error where it can then be corrected.

Now that there are no errors with the program code ensure that the PLC selected in the offline program, the one on the programming computer is the same as the actual PLC you are downloading to. For instance if you are downloading a project to a 1756-L71 ContolLogix5571 controller but the program is configured for a 1756-L71S ContolLogix5571S safety controller the program download will fail. Note too that if the firmware of the PLC is different than the revision of the offline program the download will fail as well. The controller and revision of the offline project must match the actual physical configuration of the PLC in order to download a program. It is important to note that the peripheral I/O does not necessarily have to match the actual configuration to download a program. That is to say the I/O configuration tree of the offline project may have an IA-16 Input card in slot 4 of the local chassis and the actual PLC configuration may have an OA-16 Output card in the same location. This will not stop the download, however there will be a module fault or type/mismatch fault on slot 4.

Making Online Edits

One of the nice things about using RSLogix 5000 is the power it gives you to perform online edits. Editing online is the ability to modify, delete or add code from a programming computer while online with a PLC. You also have the ability to add and delete tags.

While it is possible to create new tags and delete unused tags online you will not have the ability to change a tag's data type, for instance change a tag type from Boolean to DINT.

Online editing in some cases will be the only way for you to modify a program. This can be the case when a PLC is running a process that cannot be stopped for one reason or another.

Pro Tip: The search and replace functionality in the ladder editor is disabled when online with a program. I had a job that required me to add a couple of conveyor lines in a material-handling process. The system ran 24 hours a day and one of the requirements for the job was to insure the process did not stop while the upgrade was taking place. I had to do all my work online. The existing program was well written and documented. The system had over a hundred conveyors. Each individual conveyor had the same logic and the tags used the same data structure. Basically I just needed to copy one routine and create a new routine and paste the data in with a new tag name. I copied all the data from a routine named Conveyor_BC_001. I then pasted the code into notepad, the simple text editor that comes with windows, and used the search and replace functionalities in notepad to change any reference to Conveyor_BC_001 to Conveyor_NC_001. I then cut that code and pasted into my new routine and created a new tag called Conveyor_NC_001 of the data type conveyor. In all it took about five minutes to add the logic that ran the two new conveyors. That's the power of online editing. It is worthy to note that there are other tools such as import and export that I could have used, however I shy away from using these tools while making online edits as I have had software lock up when importing code to an online project.

The Online Edits Toolbar.

In order to do online editing you must familiarize yourself with the online edits toolbar. This toolbar will show up at the top of the ladder editor when you are online with a PLC and the online editor is visible (you can see code).

Everything on the toolbar will be greyed out with the exception of the "Start pending rung edits button". You can highlight a rung and click this button to start edits. You can also highlight a rung and double left click on the rung or press enter from the keyboard to start edits. It's also possible to drag a new rung down into the ladder editor the same way you would offline and drop the rung in the desired location.

Once you have a rung in edit mode you can add, delete or modify instructions and tags as necessary. Once the code is modified you can do one of two things. If you're feeling extremely confident with your modifications you can go ahead and click the "Finalize all edits in program" button and be done with it. This will finalize all edits in the entire program. Projects may contain many programs so be sure you know how many programs and how many edits have not been finalized. When you press this button a window will pop up asking you if you are sure. In this pop up window there will be a view to all the routines that have pending edits. Sometimes edits may not be assembled in the program. Perhaps the last technician did not completely assemble edits, so be careful not to assemble edits you did not make. In other words this is a quick and handy tool but exercise caution especially if more than one programmer is working on the code.

The other option is to manually go through the process of testing, accepting and finalizing edits. All options are available from the online edits toolbar.